100
THINGS TO
KNOW ABOUT

SAVING
THE PLANET

Usborne Quicklinks

For links to websites and videos where you can find out more about many of the facts in this book, and discover other ingenious ways to save the planet, go to **usborne.com/Quicklinks** and enter the keywords: **things to know about saving the planet**.

Here are some of the things you can do at the websites we recommend:

- See wildlife thriving at the site of a catastrophic nuclear disaster
- Take a virtual tour of the Doomsday Vault, the world's largest seedbank, deep in the Arctic
- Is your cheeseburger causing global warming? Find out the carbon footprint of different foods
- Learn how to build a hotel to help bees
- Dive into underwater worlds protected by laws, not borders
- See how art can draw attention to the destructive impacts of climate change

Please follow the online safety guidelines at Usborne Quicklinks. Children should be supervised online. Usborne Publishing is not responsible for the content of external websites.

100
THINGS TO
KNOW ABOUT

SAVING
THE PLANET

Written by
Rose Hall, Jerome Martin, Alice James,
Darran Stobbart, Alex Frith,
Tom Mumbray, Eddie Reynolds,
Lan Cook and Matthew Oldham

Illustrated by
Parko Polo, Dominique Byron,
Dale Edwin Murray, Federico Mariani,
Jake Williams and Ollie Hoff

Layout and design by
Jenny Offley, Lenka Hrehova, Samuel Gorham,
Tilly Kitching, Lucy Wain, Winsome D'Abreu,
Freya Harrison and Matt Preston

With expert advice from
Professor Mike Berners-Lee, Jessica Moss
and Professor Owen Lewis

Why does the planet need saving?

Planet Earth is our home – and the home of *all* known living things. But over the centuries, some human activity has done increasing harm to the planet.

The Earth is wrapped in a layer of air called the **atmosphere**. Like the roof of a greenhouse, it lets in sunlight and traps heat, keeping the planet warm.

Atmosphere

But for the last two hundred years, people have been burning lots of **fossil fuels** (coal, oil and gas) to heat homes, generate electricity, power factories and run cars.

Burning fossil fuels pumps out **carbon dioxide** (CO_2) and **methane gas** – known as **greenhouse gases**.

The more greenhouse gases are released into the atmosphere, the more HEAT the atmosphere traps.

People have known about this **greenhouse effect** for decades – but have kept on using fossil fuels. And so the Earth is getting hotter. In fact, it's becoming too hot to function as it should.

This is what scientists call the **CLIMATE CRISIS**.

The Earth is made up of many different **ecosystems** – groups of living things and their natural environment. These ecosystems are connected. They depend on each other. If rising temperatures cause damage to one part, then that will affect ALL the rest.

In polar regions, animals are struggling to find food because the heat means the sea ice is melting.

In the oceans, coral reefs are dying as the temperature of the water rises.

On land, wildfires, heatwaves, droughts, floods and storms are becoming more common.

To make matters worse, humans are also destroying animals' homes by cutting down forests to grow food and build roads.

On top of that, humans are dumping waste and spilling harmful chemicals, choking the natural world with pollution.

So the planet needs saving. But how?

Turn the page to find out.

A toolkit for saving the planet

There isn't one easy way to fix the climate crisis and the other problems facing the planet. It's going to take all sorts of ideas, approaches, technology and tools. But it can be done – and everyone has a part to play.

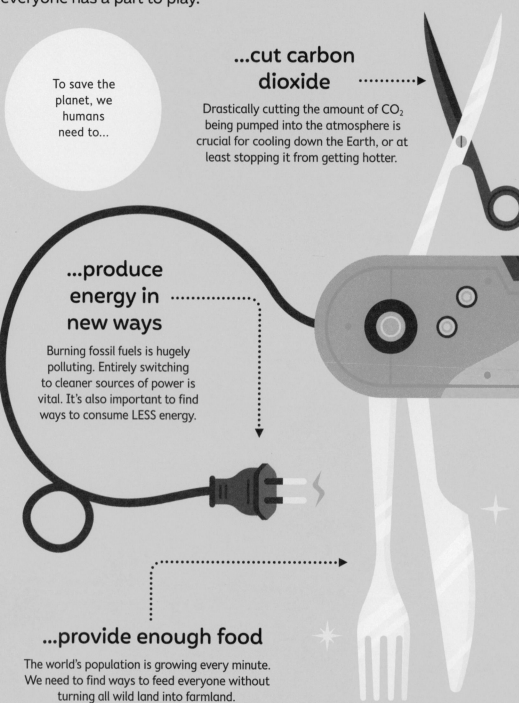

To save the planet, we humans need to...

...cut carbon dioxide

Drastically cutting the amount of CO_2 being pumped into the atmosphere is crucial for cooling down the Earth, or at least stopping it from getting hotter.

...produce energy in new ways

Burning fossil fuels is hugely polluting. Entirely switching to cleaner sources of power is vital. It's also important to find ways to consume LESS energy.

...provide enough food

The world's population is growing every minute. We need to find ways to feed everyone without turning all wild land into farmland.

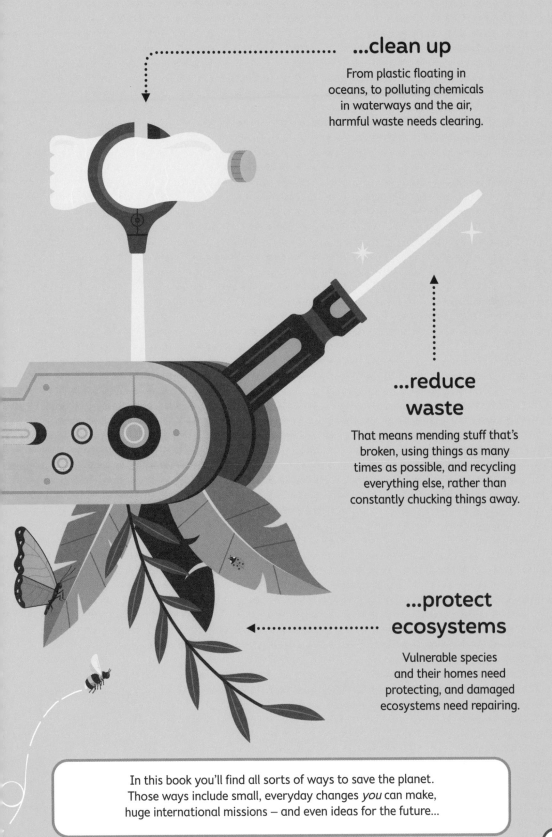

...clean up

From plastic floating in
oceans, to polluting chemicals
in waterways and the air,
harmful waste needs clearing.

...reduce waste

That means mending stuff that's
broken, using things as many
times as possible, and recycling
everything else, rather than
constantly chucking things away.

...protect ecosystems

Vulnerable species
and their homes need
protecting, and damaged
ecosystems need repairing.

In this book you'll find all sorts of ways to save the planet.
Those ways include small, everyday changes *you* can make,
huge international missions – and even ideas for the future...

1 Most of the planet's life...

is found in just 17 countries.

There are around 200 countries in the world, but just 17 of them hold most of the Earth's **biodiversity** – its great variety of different species of plants, animals and other living things. These countries are known as **megadiverse**.

Brazil, for example, home of the Amazon Rainforest, contains an estimated **4 million** species. Here are a few of them...

Brazil nut tree

Jaguar

Squirrel monkey

Cacao tree

Heliconia

Vines

Barrigona palm

Poison dart frog

Caiman

Piranha

Strangler fig

The 17 megadiverse countries make up less than **10%** of land on Earth...

...but hold more than **70%** of its biodiversity.

Hummingbird

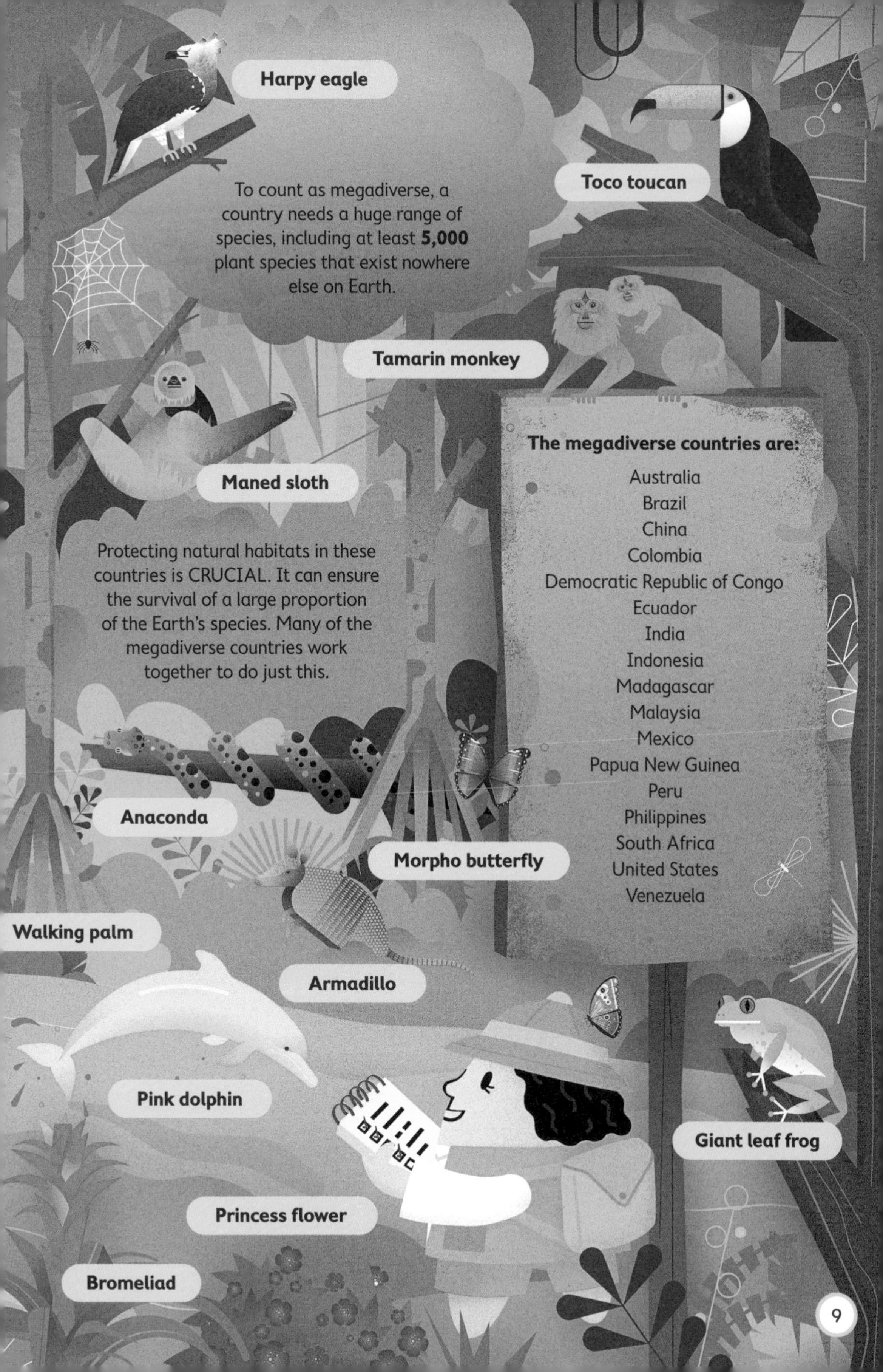

Harpy eagle

Toco toucan

To count as megadiverse, a country needs a huge range of species, including at least **5,000** plant species that exist nowhere else on Earth.

Tamarin monkey

Maned sloth

The megadiverse countries are:

Australia
Brazil
China
Colombia
Democratic Republic of Congo
Ecuador
India
Indonesia
Madagascar
Malaysia
Mexico
Papua New Guinea
Peru
Philippines
South Africa
United States
Venezuela

Protecting natural habitats in these countries is CRUCIAL. It can ensure the survival of a large proportion of the Earth's species. Many of the megadiverse countries work together to do just this.

Anaconda

Morpho butterfly

Walking palm

Armadillo

Pink dolphin

Giant leaf frog

Princess flower

Bromeliad

2 Houses could wear jeans...

to keep warm.

It can take a lot of energy to heat a home, but some of that heat often escapes. One way to save energy is to make sure that heat stays in, and that's where old jeans can come in handy.

Houses can be fitted with **insulation**, a padded material, to keep them warm.

Often, insulation is made from glass wool, produced by melting glass. But this process uses a lot of energy.

Less energy is needed to make insulation out of natural materials, such as sheep's wool or recycled jeans.

Is this how you put jeans on a house?

No! Old jeans are shredded up and pressed into panels...

...then fitted inside the walls, floor and roof.

3 Stop making microvillains...

we're already outnumbered!

Some problems are too big – AND too small – to fix. Humans have littered the planet with trillions and trillions of **microplastics**: tiny pieces of plastic waste. We may never get rid of them all, but we CAN stop adding to the problem.

Microplastics are pieces of plastic less than **5mm (0.2 inches)** across. Some are so small you need a microscope to see them.

Foolish humans! You created us, but you can never destroy us all!

Microplastics last for centuries, leak chemicals into the environment, and are often swallowed by animals – and can harm or even kill them.

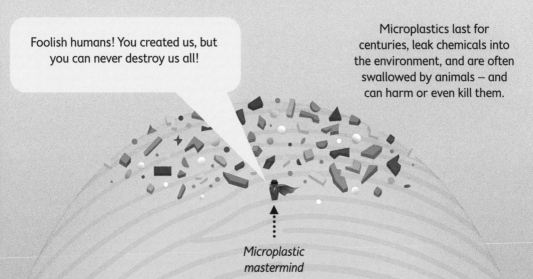

Microplastic mastermind

Microplastic pollution can be found EVERYWHERE: from the deepest ocean canyons, to seemingly untouched mountain tops, and even in our drinking water.

Lots of these microplastics come from plastic waste that breaks down into tiny pieces over time.

So one good way to combat microplastics is to reduce the amount of plastic you use and throw away, such as drinking straws, water bottles and plastic bags.

4 Squashed apple skins...

can be made into shoes.

Many shoes are made from leather as it is strong and flexible. But the process of making leather from animal skins requires chemicals which cause air and water pollution. So some businesses are striding ahead by designing planet-friendly leather alternatives.

Option 1:
use apples

After making apple juice, leftover apple cores and skins can be squashed and made into a leather-like material.

Option 2:
use pineapple leaves

The leaves of a pineapple plant normally go to waste when pineapples are harvested. But they can be processed into a leather alternative.

Option 3:
use cork bark

A leather-like material can be made from bark stripped from cork trees.

Option 4:
grow leather

Scientists can grow a substance called **collagen protein** in a lab. They shape the collagen protein into strands, which they join together to make a leather-like fabric.

Option 5:
safer process

Using rhubarb instead of toxic chemicals is a much less harmful way to turn cow skin into leather.

5 The "Roadless Rule"...

saves trees from tourists.

In 2001, the US Government introduced the **Roadless Rule**. It prevents roads being built through millions of miles of precious, protected landscapes. For example, the enormous Denali National Park in Alaska contains just one road, leaving the rest of the park, and all the life in it, untouched.

6 A *little* nudge...
can make a **BIG** difference.

One of the simplest ways to save the planet is also one of the toughest – getting people to change their habits. Here's one example – how do you persuade people to reuse shopping bags, rather than relying on stores to provide a new bag each time? The answer is to **nudge** them...

In recent years, many governments have passed laws forcing supermarkets to charge customers for plastic shopping bags. The charges are tiny, but the impact is HUGE.

Before charges were introduced in England in 2015, a typical shopper was using **140 thin plastic shopping bags** per year.

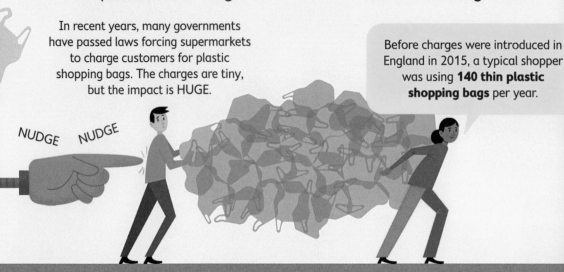

NUDGE   NUDGE

7 It's not what it's made from...
it's how often you use it.

Phasing out plastic bags is sometimes seen as a victory for the planet. But in truth, plastic makes good bags. There's a lot to think about when it comes to choosing which kind of bag might be best for the planet.

How often can the bag be reused?

Once   Over 200 times

How much energy to make the bag?

Lots and lots Not much

How long for the bag to rot away?

Many decades   A few weeks

By the end of 2016, people's habits had changed, and they were each using just **25 bags** per year...

...and by 2020 fewer still. Overall, that means less plastic being dumped in landfill sites or, worse, in the sea.

NUDGE

NUDGE

Economists call the idea behind this sort of trick **nudge theory**. Governments around the world have used a range of nudges to get people out of all sorts of habits that harm the planet.

Thin plastic	Thick plastic	Paper	Cotton

I can't wrap my head around all that! Just tell me which bag to use.

In fact, whatever it's made from, the trick is to use a bag as often as possible, and NOT to buy brand new ones.

8 The Great Horse Manure Crisis...

was solved, but replaced by another crisis.

For centuries, horses were people's main mode of transportation. But wherever horses went, they left manure. Big, stinking heaps of it. That's what led to the Great Horse Manure Crisis of the late 1800s...

In New York City, there were over 100,000 horses at work each day. They produced a lot of manure.

Layers of dung built up on the streets, attracting flies, some of which spread diseases to humans.

One day's worth of manure weighed about as much as 11 blue whales.

Oh no! I might catch typhoid!

Manure was piling up in other cities around the world, too. Nobody knew how to stop it.

Late 1800s
Horses dominated the streets of New York City.

1884
City leaders held an international conference to seek a solution, but they couldn't find one. The crisis was almost at its peak.

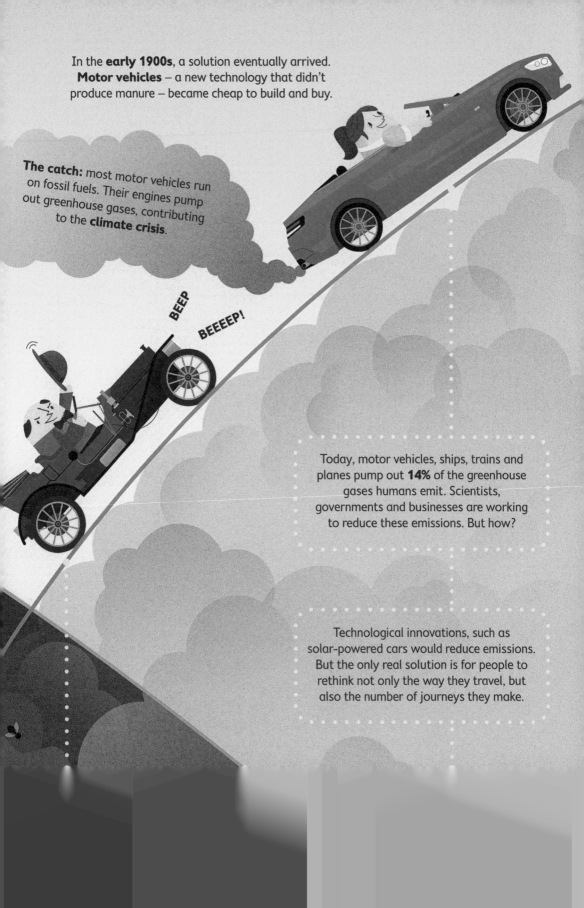

In the **early 1900s**, a solution eventually arrived. **Motor vehicles** – a new technology that didn't produce manure – became cheap to build and buy.

The catch: most motor vehicles run on fossil fuels. Their engines pump out greenhouse gases, contributing to the **climate crisis**.

BEEP

BEEEEP!

Today, motor vehicles, ships, trains and planes pump out **14%** of the greenhouse gases humans emit. Scientists, governments and businesses are working to reduce these emissions. But how?

Technological innovations, such as solar-powered cars would reduce emissions. But the only real solution is for people to rethink not only the way they travel, but also the number of journeys they make.

9 Treating animals like people...
could stop them from going extinct.

As people destroy more and more wild land, animals are losing their homes, and even facing extinction. But what if the land where the animals lived was made their property by law?

At the moment, any wild land is considered the property of *humans*. But some campaigners think *animals* should be given **property rights** – a legal claim to own the land they live on.

You need PERMISSION to build here.

GET OFF OUR LAND!

GO AWAY!

That would make it *illegal* to build on wild land without taking into account the animals' interests. These interests would be defended by a **guardian** – an independent human representative. This is similar to the way people who can't represent themselves – such as children or sick people – are given legal guardians to fight on their behalf.

In reality, this could work in a similar way to land protected as nature reserves or national parks. But it's all about stopping humans from seeing themselves as the SOLE OWNERS of the planet. It may take changing the law to change people's attitudes.

10 The Whanganui River...

is legally a person.

In New Zealand, the Whanganui River and the Te Urewera Forest both have **identity status**, meaning the law treats them as people.

These are the first natural places in the world to be protected in this way.

They can't be built on or changed. The environment, and the local Māori people's ancestral and spiritual connection to it, is preserved.

11 Slaying a vampire...

is as easy as flipping a switch.

Devices that consume electricity even when they're not in use are known as **energy vampires**. From laptops left on standby, to chargers that aren't charging anything, vampires are lurking all around us...

Energy vampires, such as TVs, computers, games consoles and devices with a digital clock, constantly gobble up electricity when they're on standby.

Devices on standby can account for **more than 10%** of a home's electricity use.

Fortunately, you can stop this by simply switching off or unplugging devices when they're not in use.

12 Eating lionfish fritters...

could stop an invasion.

Lionfish are reef-dwelling fish. Originally from the Indian and Pacific Oceans, they have invaded the Caribbean Sea in huge numbers, where they threaten to crowd out other animals. One response may be to put lionfish on the menu.

First brought to the Caribbean by humans, lionfish are now an **invasive species**: newcomers that cause lots of damage to the environment.

They breed very quickly, and are aggressive hunters, able to eat or crowd out the creatures usually found in Caribbean waters.

Luckily, lionfish are also delicious, so some scientists are encouraging people to catch them and make them part of local cuisine.

Eeeek! Save me!

Turning invaders into fritters, stews and patties could help keep their numbers down, and preserve the fragile balance of creatures in the wild.

13 A walking bus...

keeps cars off the roads.

Many children get dropped off at school in cars, but this creates CO_2 emissions and air pollution around schools. To reduce car traffic, some children walk to school in a group instead, known as a **walking bus**.

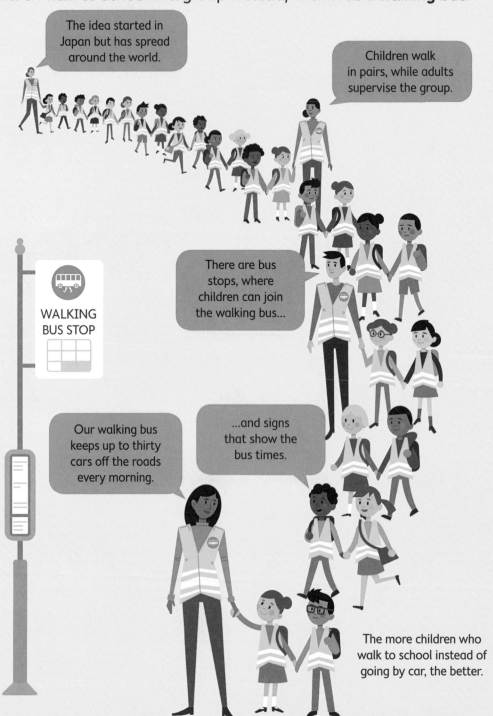

14 Save the forests...

to save the seas.

Forests are a type of **ecosystem** – a community of living things and their environment. When one ecosystem is protected, others nearby benefit too.

Here's an example of how several ecosystems form an interconnected system, called a **biosphere**.

Coast

Sea

Mangrove swamp

There's food at the shore for hungry birds.

Driftwood makes a perch for seabirds and a shelter for fish eggs.

Branching tree roots filter water.

Bird droppings fertilize soil.

Cliffs

Phosphorus found in desert dust can help fertilize trees.

Forest

Water enriches land and helps thirsty creatures.

Grasses stabilize soil and stop deserts from spreading.

River

Desert

Grassland

Turning biospheres into **protected nature reserves** is now a top priority for international organizations such as the United Nations.

23

15 Whiter walls...

make "greener" houses.

For thousands of years, buildings in hot places have been painted white to reflect the warm rays of the Sun. Now this traditional technique is being more widely used to reduce the need for air conditioning, in a movement called **cool-roofing**.

Air conditioning uses a huge amount of energy. It also releases a chemical called **hydrofluorocarbon** into the atmosphere. This is a particularly harmful greenhouse gas.

Painting a building white can reduce the temperature inside by up to 5°C (9°F)...

...enough to reduce air conditioning use by 40%.

16 Mushrooms and milkweed...

could mop up oil spills.

Oil spills from ships, pipelines and oil rigs pollute oceans and harm animals, so it's important they're cleaned up properly. Here's how mushrooms and milkweed plants could play a part...

A common clean-up method is to break oil spills into smaller droplets, using chemicals known as **dispersants**. But these are toxic, meaning they can poison wildlife.

Damaged oil rig

Researchers think oyster mushrooms could provide a *non-toxic* solution. They release substances, known as **enzymes**, that break down oil and disperse it harmlessly.

Another possibility is to use milkweed to absorb oil from the ocean's surface.

Milkweed seeds have very fine, hollow hairs, which act like drinking straws to suck up oil.

In tests, giant socks filled with milkweed hairs have been found to soak up oil twice as fast as plastic tools currently used to absorb oil spills.

17 Knitted sweaters...

protect penguins against oil spills.

Scientists working on Phillip Island in Australia have found an unusual way to help their resident colony of little penguins, or fairy penguins. The penguins are very vulnerable to oil pollution, so the scientists have taken up... knitting.

Oil leaking out of ships and through pipes sticks to the penguins' feathers. When they try to lick it off they are poisoned and die.

To prevent this, in the event of a big oil spill, penguins are given knitted sweaters. The wool traps the oil, and stops it getting onto the penguins' feathers.

The sweaters can be washed, and the penguins bathed in a clinic on the island. Once the oil spill has been cleaned up, the penguins are safe to return to the ocean.

18 It'll be "game over"...

if we don't stop burning fossil fuels.

Fossil fuels are substances made from the long, *long* dead remains of living things. When they're burned, they release energy, but also emit CO_2 Climate scientists warn that the Earth can only cope with a limited amour of CO_2 in the atmosphere – and we're rapidly approaching this limit...

The most common fossil fuels are coal, oil and natural gas.

When they are burned, they all emit CO_2 but coal produces twice as much as natural gas. Here's how their CO_2 emissions compare:

Coal

Oil

Natural gas

Imagine this grid is the atmosphere slowly filling up with CO_2. If the CO_2 reached the dotted line, the planet would become so hot that life on Earth will struggle to survive. In other words, it would mean...

...GAME OVER

Keeping the amount of CO_2 in the atmosphere as low as possible is a planet-saving priority.

Switching from coal to gas will keep us in the game for a little longer – but the only way to win is to STOP BURNING FOSSIL FUELS AT ALL.

Light beams and air jets...

pick out plastic for recycling.

Recycling waste plastic into new things reduces the amount that gets dumped. But there are lots of different types of plastic, so before anything can be recycled, it must be sorted.

This machine picks out a type of plastic called **PET**.

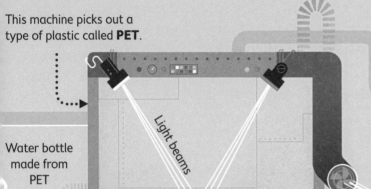

Water bottle made from PET

Light beams

Conveyor belt

A sensor detects light bouncing off PET plastic, which triggers a jet of air.

PET plastic shoots into a bin.

Detergent bottle made from HDPE

Milk bottle, made from another type of plastic, called HDPE

Other types of plastic don't set off the sensor. They are picked out by another machine further down the conveyor belt.

Black plastic is not picked out, because the light beams in these sorting machines don't bounce off black material.

Some people want black plastic to be banned because it is too hard to recycle.

Food trays made from black plastic

20 Imaginary fences...

protect real places.

Governments and international organizations can make laws to protect stretches of land and ocean. These areas may not have visible edges or boundaries, but they can shield wildlife from harm.

This page shows part of the Ross Sea Marine Protected Area, off the coast of Antarctica.

It was created in 2018 and is one of the world's largest protected areas, covering

1.5 million km²
(600,000 square miles)

– an area three times the size of France.

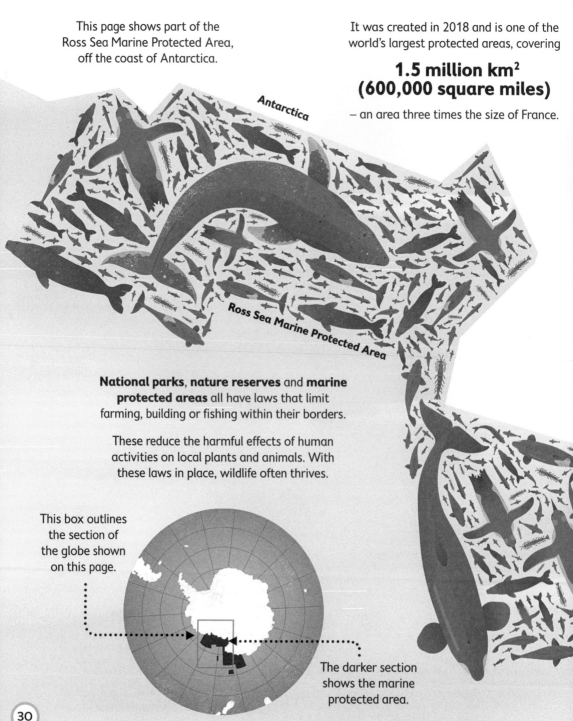

Antarctica

Ross Sea Marine Protected Area

National parks, **nature reserves** and **marine protected areas** all have laws that limit farming, building or fishing within their borders.

These reduce the harmful effects of human activities on local plants and animals. With these laws in place, wildlife often thrives.

This box outlines the section of the globe shown on this page.

The darker section shows the marine protected area.

21 Wars, disasters and toxic waste...

make accidental nature reserves.

**Rocky Mountain
Arsenal National
Wildlife Refuge**
(Colorado, US)

1942: A large chemical weapons factory was built in the grasslands near Denver. For decades, deadly chemicals and toxic waste polluted the area, so no one could live there.

Today: The land is still unsafe for people, but it has been declared a wildlife refuge, and is home to bison, deer and bald eagles.

**Korean Demilitarized
Zone**
(Between North Korea
and South Korea)

1953: To bring fighting to an end between North and South Korea, a border zone 4km (2.5 miles) wide was created. To this day, few people cross the heavily guarded zone.

Today: Rare cranes, Asian black bears and other animals make their homes among trenches, barbed wire and deadly landmines.

**Chernobyl Nuclear
Exclusion Zone**
(Ukraine)

1986: An accident at a nuclear power plant released huge amounts of deadly radiation. Over 100,000 people were evacuated from nearby cities and towns, never to return.

Today: Animals thrive and forests grow. Endangered horses, black bears and packs of wolves roam through abandoned towns.

All of these events were human catastrophes – but they also show how, despite the dangers these environments still hold, nature can recover and flourish when it is protected from human activity.

22 Hardworking earthworms...

need farmers to treat them well.

Soil may look dull, but actually it's a lively ecosystem, kept healthy by earthworms. But they can't do this useful job if farmers spray chemicals or use tilling machines to break up the soil.

We're surface-dwelling earthworms. We eat dead leaves and release **nutrients**, which plants need to grow.

WE WORK HARD – DON'T SPRAY US WITH CHEMICALS!

We're topsoil earthworms and we eat soil and excrete nutrients back into it.

We also clump small pieces of soil together, which stops rain from washing soil away.

23 Turtle-friendly lighting...

helps hatchlings see the Moon.

Baby turtles that hatch on beaches at night instinctively follow the brightest light nearby to find the sea. For millions of years, that was the Moon. But today, other lights are brighter...

Turtles follow the glaring lights of shops, bars and streets behind beaches, and end up run-over or stranded.

SHOP

LIVE MUSIC

HIGHWAY

TURTLE BEACH BAR

BAR

HOTEL

ICE CREAM

LEAVE **PLANT STALKS** IN THE GROUND AFTER HARVEST - **THAT'S OUR FOOD**.

We are deep burrowing earthworms and we mix the soil.

Our burrows make space for plant roots to grow, and help air and water flow through the soil.

KEEP SOIL COVERED WITH PLANTS ALL YEAR ROUND.

NO TO TILLING!

WE MIX SOIL FOR YOU.

When farmers stop tilling, cut chemical use and leave soil covered with plants, earthworms can continue their work maintaining the soil.

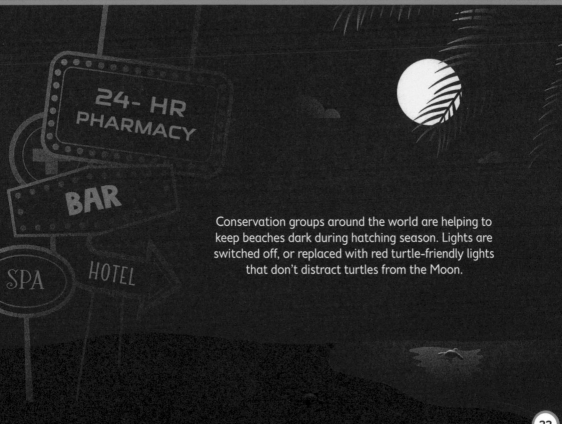

24- HR PHARMACY

BAR

SPA   HOTEL

Conservation groups around the world are helping to keep beaches dark during hatching season. Lights are switched off, or replaced with red turtle-friendly lights that don't distract turtles from the Moon.

24 Cutting down trees...

can sometimes help cut down carbon.

Trees take in CO_2 through their leaves. Most scientists agree that planting trees will help save the planet by reducing CO_2 in the atmosphere. But strange as it might sound, cutting down trees might help too...

Trees take in most CO_2 during what scientists call their **fast-growth stage**. The carbon is then stored in the trunks, branches and leaves.

Once fully grown, trees take in a lot less. That means it might sometimes be better to cut them down and plant new ones, than to leave forests to mature.

The timber could also be used as a building material, or to make products that might otherwise be made from plastic.

25 The oldest surfboard...

is still the greenest surfboard.

For centuries, surfers used boards carved from wood. Then, new materials, such as plastic foams and resins, transformed surfing, making strong, super-light boards. But today, wood is coming back into fashion...

Pre-20th century

Most surfers use solid wooden boards, known as alaias (ah-LIE-ahs) in Hawai'i, where modern surfing has its roots.

20th century

Most surfers use boards made from artificial materials, such as a plastic foam core covered with artificial resin to make it waterproof.

Once discarded, these boards take centuries to break down.

21st century

Surfers looking for greener boards rediscover alaias.

They are made from nothing more than planks of wood, rubbed with seed oil to protect them from salt water. Once discarded, they quickly rot away.

And it's not just sports equipment – it's everyday items such as toothbrushes, t-shirts and packaging. By choosing products made from **biodegradable** materials – materials that break down safely and naturally after use – people are reducing their impact on the planet.

26 Face masks for cows...

could clean up their burps.

When cows digest food such as grass, corn or soy beans, they burp out a greenhouse gas called **methane**. The problem has become so bad that scientists are developing face masks and food supplements that could make cows' burps less harmful.

There are a billion of us cows around the world, so the methane from our burps heats up the planet a lot.

BURP

BURP

BURP

BURP

Well, instead of that sparkly mask, you should get one like we have. Ours convert methane in our burps into CO_2 and **water**.

CO_2 and water are also greenhouse gases, but they trap less heat than methane, so overall they're less harmful.

Scientists are also experimenting with changing cows' diets to reduce the volume of methane they produce.

COW SNACK SHACK

BURP

GRASS with METHANE-MINIMIZING toppings...

A SPRINKLE OF SEAWEED

or

A DRIZZLE OF OIL

Forget face masks... Because I eat seaweed with my food, I don't produce much methane at all.

I like linseed oil with my meals. It helps my digestion, so I burp less methane.

COW SNACK SHACK

Methane isn't the only problem with cattle farming. See pages 49 and 84-85 to find out more.

could save dying reefs.

Introduction: Pollution, rising water temperature and disease are causing the world's coral reefs to die. This could be catastrophic - reefs are home to **25% of all sea life**.

Hypothesis: Special lab-grown corals may be one solution to the problem by helping to build reefs that can survive rising water temperatures.

Experiment: Breed two or more different types of coral to create **hybrid corals**.

Method: Corals are grown in tanks where the water conditions can be adjusted to mimic various ocean environments.

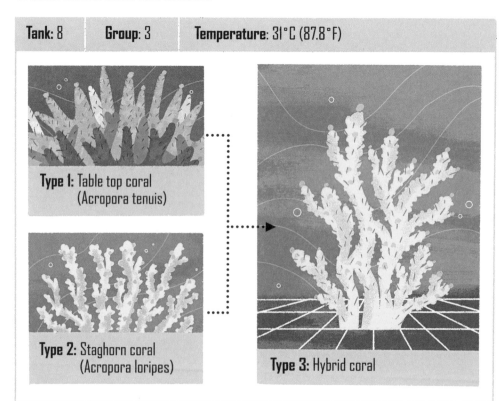

| Tank: 8 | Group: 3 | Temperature: 31°C (87.8°F) |

Type 1: Table top coral (Acropora tenuis)

Type 2: Staghorn coral (Acropora loripes)

Type 3: Hybrid coral

Results: Many hybrid corals show much greater resilience to rising water temperatures.

Conclusion: After three months, hybrids are placed in the wild to create reefs with a much greater chance of survival.

needs cleaning up.

Everywhere people explore, they leave stuff behind – from astronauts' junk left on the Moon, to the robots surveying Mars.

On Mars...
there are already at least

12

different broken-down machines just sitting there.

Leave me alone!

On the Moon...

over **750**

separate items have been left behind, from used wet wipes to abandoned vehicles – enough to fill ten household waste collection trucks.

Clean up your mess, Earthlings!

In Earth's orbit...
there are more than

128,000,000

pieces of junk

from old satellites, to stray bolts from spacecraft.

How do you think I feel?! Every time a rocket blasts into space, its used-up parts are ditched into orbit. Humans are taking pollution to the next level.

The European Space Agency is currently working on systems to remove space junk from Earth's orbit...

...to prevent damage to working spacecraft, and to protect Earth's environment.

29 More whale dung...

would cool the planet.

Whales eat deep underwater, where they hunt animals such as squid or krill – but they can only excrete their dung at the surface. And that's a good thing for the planet.

Nutrients such as **iron, phosphorus** and **nitrogen** can be scarce at the ocean's surface...

Whale dung (full of nutrients)

...but whale dung is full of these nutrients, so when it spreads through the surface water, it feeds millions of tiny living things called **phytoplankton.**

Those phytoplankton suck CO_2 out of the atmosphere and store the carbon as they grow.

In fact, the phytoplankton in oceans around the world store as much carbon as about **1.7 trillion trees** – roughly four Amazon rainforests.

If we made the oceans safer for whales, there would be more of them – and more whale dung. The number of phytoplankton would jump, which would mean less carbon in the atmosphere and a cooler Earth.

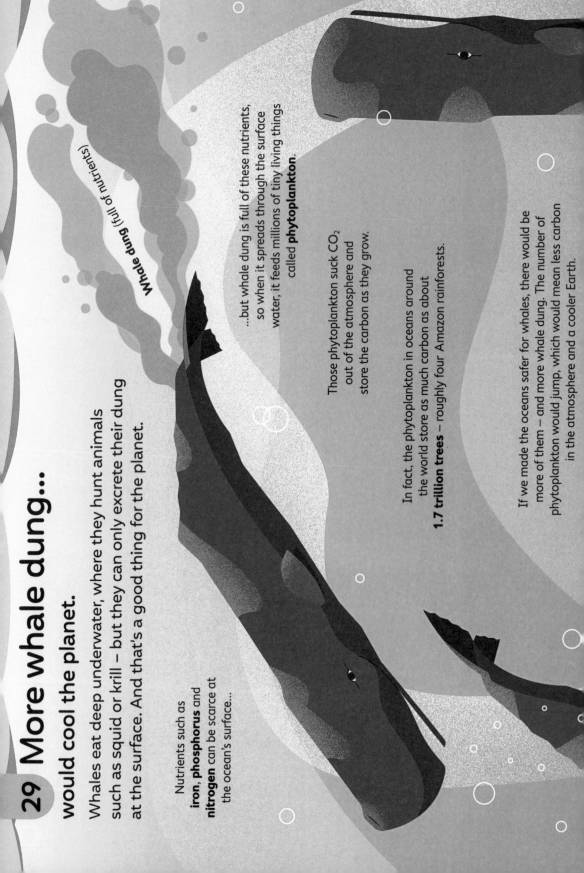

30 A living whale is worth...

two million dollars.

Money is a powerful tool. If enough money is at stake, organizations and governments WILL act.

That's why a team of economists has calculated the value of a single living whale. They hope that, if everyone realizes just HOW MUCH whales are worth, people will take greater steps to protect these vulnerable creatures.

Some of the jobs whales do for us (for free):

- Reduce carbon by excreting at the surface of the ocean
- Increase fish populations by feeding phytoplankton – a major food source for other fish
- Boost tourism by attracting whale watchers

Value per whale in US dollars:

$2,000,000

Total value of work done by whales:

1 TRILLION US dollars

The up-and-down movement of whales bringing nutrients from the deep to the surface is known as the **whale pump**.

Sperm whale (full of squid)

Squid (full of nutrients)

31 Farmyard robots...

can reduce chemical pollution.

Many farmers spray chemicals over their fields to get rid of weeds. The trouble is, these **weedkillers** also kill wildlife, pollute rivers and damage the soil. But Swiss inventors have created a robot that can cut weedkiller use by 90%, by targeting weeds VERY precisely.

Cameras facing the ground take photos to help the robot identify the plants it passes.

The robot compares each photo to a library of plant photos to determine which plants are weeds.

Built-in solar panels power the robot using energy from the Sun.

Robotic arms move into position to spray a small dose of weedkiller on the weeds.

Similar robots are being designed which pinpoint weeds and remove them with blades instead of weedkillers.

Robots could be programmed to leave some harmless weeds in the soil, to support wildlife living there.

32 Bald eagles were saved...

by a single ground-breaking book.

From the 1890s to the 1950s, the number of bald eagles in the United States mysteriously fell from 100,000 to 800. But in 1962, a book by biologist and nature writer **Rachel Carson** revealed the cause: a chemical pesticide called **DDT**.

In her book, *Silent Spring*, Carson described how DDT wasn't only killing insect pests, but other wildlife too.

DDT was washing into rivers and contaminating plants and fish. When bald eagles ate the fish, DDT was building up in their bodies.

This prevented bald eagles from laying healthy eggs. Chicks didn't survive and the numbers of bald eagles dropped.

Silent Spring was a bestseller. Hundreds of thousands of people read the book, including politicians.

Other scientists came out in support of Carson, and campaigners called for DDT to be banned.

In 1972, general use of DDT was banned in the US. By 2007, there were over 18,000 bald eagles and the bird was declared no longer at risk.

Careless flushing...

created a monster.

In September 2017, a disgusting discovery was made in the sewer system of Whitechapel, London – a mass of fat, oil and waste known as a **fatberg**.

IT STINKS!   **IT SPREADS DISEASES!**   **IT'S POLLUTING!**

THE FATBERG

It weighed more than
20 elephants
130,000kg (286,600lb)

It blocked up sewers for
250m (820ft)
the length of 13 buses!

STARRING THE UNFLUSHABLES:
COOKING OIL   WET WIPES   PLASTIC BAGS   FOOD WASTE

Take care! The Whitechapel fatberg was a record-breaker at the time, but fatbergs have been found in cities around the world. They can take teams of engineers several days to clear. To fight fatbergs, take care what you flush!

34 Recycling more than ever...

isn't stopping the planet from drowning in waste.

The government of the United States has been collecting information about recycling since 1960. Half a century later, Americans are recycling 23 times more waste. This sounds good – but sadly the total amount of waste is STILL growing and growing.

US in 2017

Total amount of waste:
**243 billion kg
(536 billion lb)**

US in 1960

Total amount of waste:
**80 billion kg
(176 billion lb)**

Total amount recycled:
**116 billion kg
(256 billion lb)**

Total amount recycled:
**5 billion kg
(11 billion lb)**

And this is only talking about the US – if we consider the rest of the world, it's even worse. Recycling old unwanted things is a good thing to do. But to stop our planet from being deluged with waste, we need to REDUCE the amount of stuff we buy, and REUSE it as much as possible.

35 Seeds and shovels...

could save the world.

It doesn't take much to plant a tree. A seed, a small hole in the soil, some sunshine, a little rain... and then some time for it to grow. But some scientists now think if we planted enough trees, it could make a significant difference.

Just as we breathe in oxygen, trees absorb CO_2 from the air. They take it in through their leaves, and they use it to grow. The CO_2 taken from the atmosphere is **stored** inside the tree as it grows.

Using satellite images, scientists have calculated that there are **0.9 billion hectares (3.5 million square miles)** of the Earth's surface that *could* be planted with trees. That's an area almost the size of the United States.

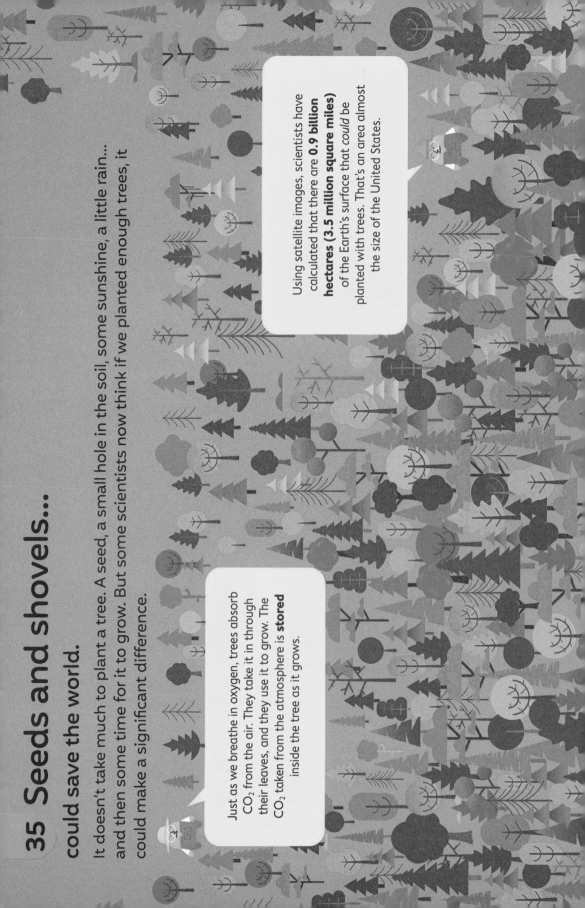

36 Jeans libraries...

make the most of every pair.

Making a new pair of jeans produces CO_2 and uses a lot of water. So it's better for the planet if fewer pairs are made, and instead each pair is worn more often. One way to do this is through a **jeans library**.

After joining a jeans library, customers can rent out a pair of jeans for weeks, months or even more than a year.

When customers are ready for a change, they bring back the jeans and exchange them for a different pair.

Here's a bigger size.

I'm bored with these.

Would you like to try a different style this time?

I've grown out of these.

These jeans are ripped.

We'll wash and repair them before renting them out again.

This system results in each pair of jeans being worn many more times than they would otherwise.

When the jeans fall apart, the jeans library recycles the fabric.

Similar libraries are in operation, where tools, toys, luggage and many other things can be rented.

37 Making just one burger...

uses more water than you drink in three years.

Farming cows for beef, and growing crops to feed them, takes a lot of water. But there's a limited amount of fresh water to go around, so going without a burger is one way of saving a surprising amount.

If you saved all the water that went into producing the beef for one burger, you would have enough for...

...all your showers for

2 months

That's if you usually shower for 5 minutes a day.

...all your toilet flushes for

4 months

That's if you flush it four times a day.

...all of your drinking water for

3 years, 4 months

That's if you usually drink around 8 glasses a day.

The amount of water it takes to produce something is known as its **water footprint**. Turn to page 52 to find out more.

Our solar system is located inside the Milky Way galaxy – a vast, swirling disk made up of dust, planets and around **100 billion stars**. But, in some places on Planet Earth, the galaxy seems to have vanished from the sky. The cause? Light pollution.

Light pollution comes from artificial light, such as street lighting, that shines beyond where it is needed, blotting out natural darkness.

This haze of light, hovering over a city and blocking out the stars, is called **skyglow**.

Caused by the glare, reflection and scattering of excess light from below, skyglow can stretch for hundreds of miles *beyond* the city limits.

The light dazzles migrating birds and leads them astray. And, due to light pollution, **one third** of people in the world can't see the Milky Way.

Much of this light isn't needed at all. It shines on empty buildings and in near-deserted streets.

Every 24 hours in the US alone, wasted lighting uses up enough energy to heat and power over **five million households** for a day.

This is what the Milky Way looks like from Earth – *without* light pollution.

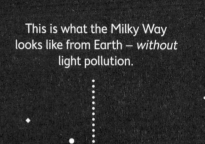

So, by dimming or turning off outdoor lights, we can save enormous amounts of energy, allow animals to go about their business...

...AND we could also enjoy the sight of a galaxy full of stars.

39 Every shoe...

has three footprints.

As well as the pattern left by its sole, every shoe has two extra footprints: a **carbon footprint** and a **water footprint**.

Anything that causes greenhouse gas emissions is described as having a **carbon footprint**.

Here are some of the things that give shoes a carbon footprint:

Shoe factories run on electricity. This comes from power stations that make greenhouse gases.

Shoe leather comes from cows, which breathe out lots of greenhouse gases.

Shoe parts must be shipped to factories, and finished shoes are shipped all over the world. Ships burn fuel, which creates more greenhouse gases.

An object's **water footprint** describes how much clean water was used to make, transport and sell it.

The impact on our planet of buying a pair of leather shoes is similar to driving 60km (37 miles) in a car.

A shoe's water footprint comes from:

Rearing cows for leather

Generating electricity

Making paper for shoe boxes

One way to reduce the world's total of BOTH footprints is for everybody to make, sell and buy **less stuff**.

40 Even phone calls...
have footprints.

It's not just physical things such as shoes, or phones, that have a carbon footprint. Non-physical things, such as phone calls, emails or internet searches, also have an impact on the environment.

Here's how...

Signals from each phone are sent around the world by base stations, which run on electricity...

Power for the phone comes from electricity...

Calls are connected by machines known as switchboards, which run on electricity...

A phone company needs buildings and people and machines, which all use up electricity...

...and that electricity comes from power stations that emit greenhouse gases into the atmosphere.

Don't feel bad about making calls!

The worst offenders are companies who make unwanted calls.

These **"junk"** messages – often called **"spam"** – make up around **30%** of all phone calls.

41 Bus stops for bees...

increase biodiversity.

Bee habitats around the world are being destroyed, causing bee populations to fall. To help solve this problem, a Dutch city has turned over 300 bus stops into bee sanctuaries.

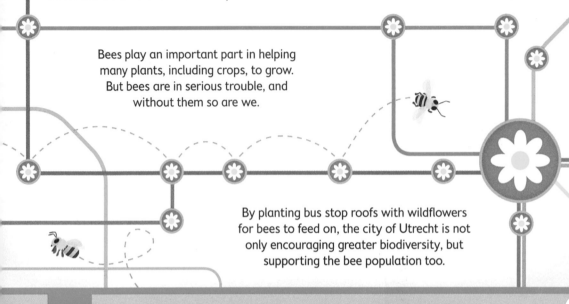

Bees play an important part in helping many plants, including crops, to grow. But bees are in serious trouble, and without them so are we.

By planting bus stop roofs with wildflowers for bees to feed on, the city of Utrecht is not only encouraging greater biodiversity, but supporting the bee population too.

42 Collecting plastic bottles...

can get you a free ride.

As more and more mounds of waste plastic build up, governments and businesses are finding ways to encourage people to recycle. In one Indonesian city, people can pay for bus rides using plastic bottles.

= 1 ride

This bus collects **250kg (550lb)** of plastic for recycling a day.

Around the world, similar systems have been introduced, including ones using plastic bottles to pay school fees or for meals. Rewards like these are known as **incentives**.

43 Powerful art...

can make people care.

The climate crisis is in the news a lot. But sometimes people have a stronger emotional response to the message when it is represented in art.

In 2017, the artist **Lorenzo Quinn** created a sculpture in Venice, Italy, about the climate crisis, called *Support*.

It makes me concerned about rising sea levels. What about you?

44 It's not what's possible...

it's what's not *impossible.*

There's a colossal amount of ice at the Earth's poles. As that ice melts, the seas swell, and places at sea level are in danger of being flooded. In a desperate attempt to protect the ice, and the animals that live on it, scientists are having to explore some ambitious ideas.

Or we could pump seawater back onto the glacier and freeze that?

Antarctica

Thwaites Glacier

Southern Ocean

What if we... built a wall?

Why?

Warm, salty water flows at the bottom of the Southern Ocean. An **underwater wall** could stop it from reaching the ice and melting it.

Wall

But to work, the wall might have to be to be over 300m (985ft) tall, and 120km (75 miles) long! That would be so expensive to build!

Lots of ideas are possible in *theory*, but in *practice* would cost too much or use far too much energy. No single solution is entirely perfect, but a choice has to be made.

45 Burning trees in the rainforest...

kept ancient farms fertile.

The Amazon Rainforest is full of trees and plants, but the soil isn't fertile enough to grow crops year after year. Since ancient times, people have solved this problem with a substance known as *terra preta*, or **black earth.**

Around 3,000 years ago, people living in the Amazon started building bigger settlements.

They cut down trees and burned the wood, which produced charcoal.

The Amazonians mixed the charcoal into the soil to produce black earth. They found that this helped their crops to grow.

Farming can make soil less fertile over time. But adding charcoal helps soil hold on to nutrients, microbes and water that keep it fertile.

As the planet's population continues to grow, we'll need to make sure there's enough food for everyone to eat without making the soil less fertile. Scientists think a modern version of black earth, known as **biochar**, could help farmers do just that...

46 Burning dung...

keeps carbon out of the atmosphere.

Leaving farm waste to rot or throwing it onto a bonfire releases CO_2 into the atmosphere. It's much better for the planet if farmers burn their waste in specially designed kilns to create **biochar**.

Farms produce lots of waste, such as manure, wood, or scraps left over after harvests.

This waste can be burned in a kiln without any oxygen...

...to produce biochar which is packed full of carbon.

Biochar can be mixed into soil, where it fertilizes the land. The carbon in it then remains buried, or **sequestered**, for hundreds of years.

47 Busy beavers...

can hold back a flood.

The climate crisis means that floods are becoming more common, as heavier rainfall causes rivers to overflow. But there is a hardworking creature which can help protect towns and villages by building flood barriers: it's the beaver.

Beavers are common in North America, but died out in the UK around 400 years ago.

Beavers have recently been released back into the wild in some parts of the UK, with positive effects...

We gnaw down trees and use wood to build a barrier, called a **dam**.

Water backs up behind the dam, creating a deep pond.

Without the dam, heavy rain would rush down the river all at once and flood human homes and farms downstream.

Because our dam is built from sticks, it's slightly leaky. It holds back the extra water and prevents flooding, without completely stopping the flow of the river.

The pond keeps the entrance to the beavers' home underwater, so foxes can't get inside.

Our new pond also boosts biodiversity, because it provides a home for a wide variety of wildlife.

Dragonfly

Kingfisher

Butterflies

Water voles

Bat

Brown trout

We aren't always welcome. But if our pond floods a field, farmers can adjust our dam so less water backs up behind it.

Frog

And farmers can put up fences around trees to stop us from gnawing them down.

48 Just one new scarlet robe...

can crowd a landfill site with junk.

We all tend to *want* new things, but *having* them isn't always good for us – or the planet. This is what Denis Diderot, a French philosopher, found 250 years ago: one new thing can quickly lead to another... and another... and another.

It all began when Diderot bought himself a beautiful new scarlet robe.

He noticed that his old chair looked shabby next to his new robe – so he bought a fine leather chair... and threw out the old one.

He replaced some old paintings with new ones that matched the chair...

...and bought a new clock to fill a gap on the wall...

...and, before he knew it, his home was filled with things that looked great together...

...but which he hadn't previously wanted and didn't actually need.

"I was the master of my old robe – but I have become a slave to my new one," he wrote.

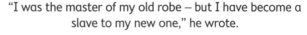

Today, economists refer to the **Diderot Effect** to explain the way people fall into the trap of buying more and more new products. This adds more and more stuff to our landfill sites – a trend that must be stopped for the sake of the planet.

49 Scientists recruit animals...

to gather planet-saving data.

There are many ways animals can help scientists gather data about our planet. Here are just a few.

Scientists can fit blackbirds with tiny transmitters that track the birds' locations. They use this information to see how climate change affects migration patterns.

Endangered species, such as jaguars and sea turtles, can be tracked so scientists can learn more about where they live, and how to protect them.

Far-flying albatrosses can carry GPS sensors to detect radar signals from illegal fishing boats. This makes the boats easier to locate, so that authorities can stop them.

Seals in Antarctica can be equipped with sensors that record the temperature of the ocean. This allows scientists to monitor where water is warming fastest, and predict when large chunks of sea ice might melt.

50 All the world's energy...

could come from a fraction of a desert.

In just one year, enough sunlight hits the Earth to meet current energy demands for more than **80 years**. But electricity generated from sunlight, known as **solar power**, currently makes up less than **5%** of the power used around the world. What if that could change?

The Sahara is one of the largest deserts on Earth, but more importantly, it's one of the planet's sunniest places.

Based on current estimates, covering less than **2%** of the desert in solar panels — which convert solar energy into electricity — could provide enough energy for the whole world.

To give a sense of what that could look like, these solar panels cover 2% of this page.

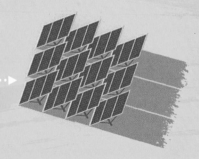

So what are we waiting for? 2% sounds small, but it would still mean covering thousands of miles of desert in solar panels.

That wouldn't be easy, and it's currently impossible to transport the energy around the world. BUT this shows the great potential of solar power, and finding new ways to harness it will be crucial.

Even on a bright sunny day, only around half the Sun's energy reaches Earth through the atmosphere. So would it be better if we put solar panels up in space?

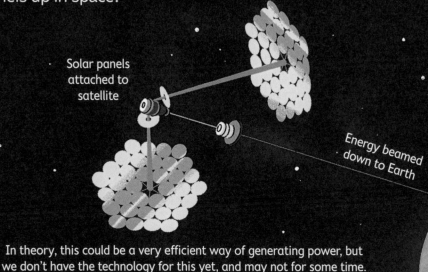

Solar panels attached to satellite

Energy beamed down to Earth

In theory, this could be a very efficient way of generating power, but we don't have the technology for this yet, and may not for some time.

‗ ☐ ✕

52 From: me...

To: you.

Subject: Saving the planet.

Typing an email and sending it might seem quick and simple, but even emails have a carbon footprint (see page 52).

Whenever a phone, computer or device is on, it is using electricity. That means every email you send or receive is keeping you on a device for longer, and using more electricity.

The worst offenders are companies who send unwanted, or junk emails. These emails – known as "spam" – make up over 50% of ALL emails sent. Cutting down on these could cut a huge amount of carbon.

Thank you.

53 Flying sausages...

turn greedy creatures into picky eaters.

An Australian mammal, the **northern quoll**, is at risk of dying out. That's because quolls are gobbling up **cane toads**, which poison and kill them. So **conservationists** are teaching them to be more picky about their food.

Cane toads are what's known as an **invasive species**. Originally from Latin America, they were brought to Australia in the 1930s and spread rapidly, posing a major threat to local quolls.

Cane toads

Quolls eat fruit, insects, lizards and frogs – they're not supposed to eat cane toads.

Northern quolls

Eating a single cane toad would kill a northern quoll.

To try to save northern quolls, scientists are making special sausages, and flinging them from helicopters into the remote areas where quolls live.

The sausages are made from a type of toad meat that tastes like cane toads, but isn't poisonous.

Scientists have also added a chemical that makes quolls feel nauseous for a short time after eating the sausages.

So, a quoll learns to associate the smell of toad meat with a horrible experience.

Next time it sniffs a cane toad, it doesn't feel like eating it, so isn't poisoned.

54 Buying wonky fruit...

teaches supermarkets a lesson.

Most supermarket displays make it seem as if each kind of fruit and vegetable has just one shape. But, in reality, they grow in all sorts of shapes, from completely straight to totally wonky. Sadly, wonky fruit is often thrown away.

Check me out.

Pick me!

We are beautiful, no matter what you say.

I'm tasty on the inside.

Many supermarkets refuse to buy wonky fruit from farmers — they say that customers prefer to buy "normal-looking" fruit.

Appearance is one of the reasons why up to

one third

of all food is thrown away. This is a massive waste of the water, land and energy used to farm it in the first place.

But there is a way to help. Choose wonky fruit whenever you can, so that supermarkets get the message that customers will buy it, and it won't go to waste.

55 Building a mountain range...

could cool down the planet.

When rocks are exposed to wind and rain, they slowly wear away in a process called weathering. As this happens, CO_2 in the air binds to the rock, and is removed from the atmosphere.

Weathering absorbs **one trillion kg (two trillion lb)** of CO_2 a year, but that's a tiny fraction of what is emitted.

Adding more rock to Earth's surface could help stop CO_2 building up and warming the planet.

This idea is known as **enhanced weathering**.

Scientists think this may have happened naturally millions of years ago, when the Himalayan Mountains formed. The new mountain range absorbed so much CO_2 that temperatures dropped by a huge **8°C (14°F)**.

Of course, we can't just build a new mountain range...

...but scientists are working on ways to add rocks to land, to put this theory into action.

HEAVE

GRUNT

56 Better golfers...
leave cleaner oceans.

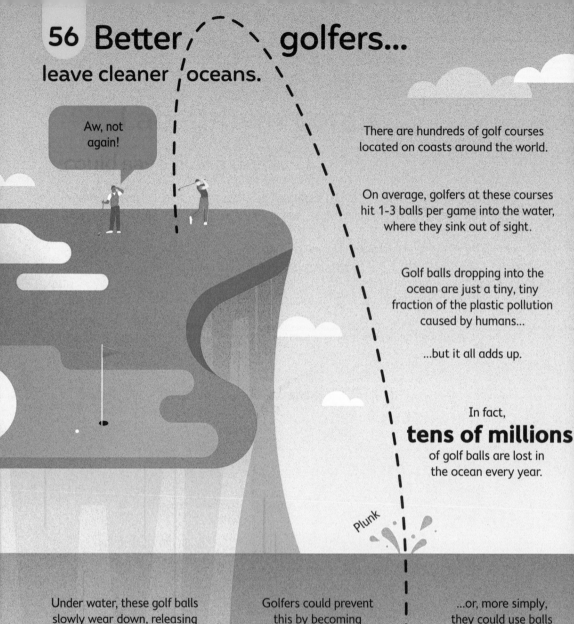

Aw, not again!

There are hundreds of golf courses located on coasts around the world.

On average, golfers at these courses hit 1-3 balls per game into the water, where they sink out of sight.

Golf balls dropping into the ocean are just a tiny, tiny fraction of the plastic pollution caused by humans...

...but it all adds up.

In fact,

tens of millions
of golf balls are lost in the ocean every year.

Plunk

Under water, these golf balls slowly wear down, releasing tiny pieces of plastic, rubber, and zinc, which can be harmful to marine life.

Golfers could prevent this by becoming better at golf, and keeping balls out of the ocean...

...or, more simply, they could use balls designed to break down harmlessly in water.

57 To solve some problems...

you have to dive in and take action.

People wouldn't know nearly as much about all the golf balls polluting the oceans if not for the efforts of **Alex Weber**, a teenager from California.

In 2016, while diving on the ocean floor, Alex discovered huge numbers of golf balls. She decided to do something about it.

With her friends and family, she collected **tens of thousands** of golf balls...

...but even then, there were still many more balls left in the ocean.

So Alex contacted a scientist at Stanford University who researches ocean pollution, and told him what she had found.

Together, they wrote and published a scientific paper explaining how lost golf balls are harming the ocean.

Their work has inspired both golfers and ocean conservation groups to start tackling this hidden source of pollution.

58 Beech today...

oak tomorrow.

Planting trees contributes enormously to absorbing and storing carbon. But to make a significant difference, it has to be the *right* trees in the *right* places – and, most importantly, a mix of *different* trees.

In different environments, scientists need to choose the types of trees that will grow best. These are often types that have been growing there a very long time.

For example...

In icy parts of Russia, it's best to plant pine trees that can survive freezing winters.

In the Amazon Rainforest, the best options are trees with deep roots that grow to huge heights, such as mahogany.

On the mountainsides of Lebanon and North Africa, cedar trees that grow well in dry, rocky conditions are ideal.

Pine

Mahogany

Cedar

Fig

Scientists need to think ahead when they do this. It's not just about what conditions are like now. It's also about what conditions will be like in 50-100 years, when the tree has grown and is mature – because that's when a tree will be storing the most carbon.

For example, beech trees currently grow well in many parts of England. But they don't cope well with drought.

As the climate warms and changes in the coming decades, there will be far more droughts in England.

Beech

Oak

Ash

Maple

So planting a tree that can cope with drought, such as an oak, might be a better option.

Overall, environments are healthiest when lots of different trees are planted together. Areas with a wide range of tree species can absorb more carbon and support different animal species.

59 Many munching mammals...

could keep the ground in Siberia frozen.

Soil that stays frozen all year round is known as **permafrost**. A layer of permafrost in Siberia, Russia, is beginning to melt, with potentially dangerous consequences. Scientists on a Siberian nature reserve are testing a way to stop the melting, using herds of large, hungry grazers.

Siberia, 1996:

I have an idea! What Siberia needs is more GRASS.

Grass? Why?

As the permafrost melts, it releases greenhouse gases, which trap heat in the atmosphere and warm up the planet.

And grass might keep the soil frozen?

Yes! Paler greens reflect more heat from the Sun than darker ones. The more grass there is, the more heat gets reflected, keeping the ground cooler.

We'll help you grow more grass!

Elk

Yak

Wild horse

Forest

Grass

Shrub

First, we'll eat the grass.

Eat it!? But we need MORE grass.

The more grass we eat, the more droppings we produce. Nutrients from our droppings fertilize the soil and help more grass to grow.

Grazers also eat other plants, making space for more grass. That will help grass spread across Siberia, keeping the permafrost cool.

60 Animal actors...

are helping to recreate a long-lost landscape.

Around 13,000 years ago, Siberia was covered almost entirely in grass, and roaming with mammals. Recently, near Chersky, Siberia, some of those mammals have been **reintroduced**. Others from that time are now **extinct**, so scientists are looking for animals that could take their place.

GRASSLAND FILM STUDIOS

Cast so far includes:
Reindeer, elks, yaks, muskoxen.
These all lived in Siberia
13,000 years ago.

Extinct animals we need actors for:

Woolly mammoth

Steppe wisent

Pleistocene camel

> What role are you auditioning for?

> Pleistocene camel. My humps store fat to give me energy during the winter, and I LOVE eating grass.

> Sounds like you stand a chance!

Bactrian camel

AUDITION ROOM

> Hi, I'm a wood bison and I'm auditioning for steppe wisent.

> Nice thick wool for the cold, and your grass-based diet is almost perfect!

> And trust me, I poop a lot! The soil will get plenty of nutrients.

DIRECTOR

Recreating a landscape by reintroducing animals is a form of conservation known as **rewilding**. There are many rewilding projects under way all around the world.

61 Discarded oyster shells...

can clean polluted waters.

Human activity is causing more and more pollution along coasts and in rivers. One possible solution is thrown out every day by restaurants around the world: oyster shells.

Rain washes polluting chemicals such as nitrogen and phosphorus off farmland into rivers and oceans.

In large quantities, these chemicals cause vast growths of algae. These algae then suck oxygen out of the water, making it impossible for other plants and animals to survive.

Some conservationists are fighting this problem by dumping oyster shells into the water. There, they sink and form a hard, stable layer called **culch**.

62 Frozen time capsules...

help predict our climate future.

In order to tackle climate change, scientists need to be able to measure it. One way is by studying **ice cores**, cylinders of ice extracted from the Antarctic and the Arctic. Here are some of the things scientists look for...

Cores are extracted vertically from deep in the polar ice.

This section is the oldest, from deep down. It shows a lot of snow was falling at the time when it formed.

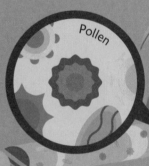
Pollen

Snowflake

Pollen shows how abundant plant life was and how long seasons lasted.

They're then cut into sections to be studied.

Over time, living oysters anchor themselves to the culch, and gradually form large groups, or colonies.

Oysters feed by pumping water through their bodies, filtering out and eating up the dirt, algae, plankton and pollutants.

Just one oyster can filter nearly a bathtub's worth of water every day, so healthy oyster colonies make for cleaner, clearer coasts and rivers.

A layer of ash in this section shows when volcanoes were erupting.

Carbon dioxide

C

More recent layers have much more CO_2 in them. CO_2 is a **greenhouse gas** that contributes to the planet becoming hotter.

Radioactive particles show there were lots of cosmic rays from the Sun hitting Earth during this time.

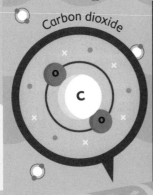

Bubbles of CO_2 trapped in the ice indicate how much of it there was in the atmosphere.

Radioactive particles

Studying changes in CO_2 levels – especially rapid rises – has provided key evidence of one of the main causes of climate change.

Ash particles

63 Furoshiki-wrapped presents...

are a gift to the planet.

Tearing wrapping paper from gifts can be exciting, but it produces lots of waste. **Furoshiki**, a traditional Japanese method of wrapping gifts in cloth, provides a planet-friendly alternative to mounds of crumpled-up paper.

An estimated **365,000km (227,000 miles)** of wrapping paper are thrown away each Christmas in the UK alone...

...that's enough to wrap around Earth's equator **9 times**. Lots of this paper is shiny or covered in glitter, so can't be recycled. But it should be reused.

Using furoshiki, a single square of cloth can be folded to wrap all sorts of gifts, over and over again.

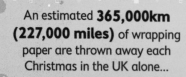

Knots hold the cloth in place, so no tape or ribbon is wasted.

When the cloth isn't being used to wrap gifts, it can be used as an alternative to a plastic bag.

Spoons you can eat...

cut down on plastic.

Every year millions of single-use plastics, such as knives, forks and spoons, are thrown out. Edible spoons may be one way to reduce the amount of waste.

Miss Freeze's Sweet Treats

Which spoon would you like?

PICK YOUR ICE CREAM

- Strawberry
- Raspberry
- Chocolate
- Vanilla
- Mint
- Toffee

PICK YOUR SPOON

Plastic

Despite the subtle taste of chemicals, carbon emissions when these are made and zero chance of being recycled, this is still our most popular spoon!

Wood

Billions of trees are cut down every year. Our wooden spoons might be compostable but they're certainly not tree-friendly!

Edible

Made from 100% natural ingredients such as rice, wheat and cornflour. In case you don't want to eat it, it's **biodegradable**, meaning it'll decompose in a few days.

Hmm...

I brought my own spoon with me.

Yum!

65 Underpasses and bridges...

help red crabs to reach the sea.

Every November on Christmas Island, in the Pacific Ocean, hundreds of thousands of red crabs emerge from the forest and make their way to the shore, to lay their eggs in the ocean. But this journey can be very risky

Australian authorities in charge of Christmas Island use all sorts of tactics to help crabs reach the sea without getting lost or killed on the way.

Many roads are closed.

A specially designed crab bridge crosses the main road.

TOOT

NO ENTRY
Crabs crossing!

Underpasses take crabs beneath dangerous roads.

That way?

Which way?

THIS way!

Miles and miles of plastic fencing funnels crabs into the right areas.

The local radio station broadcasts updates to residents.

"Latest CRAB NEWS. It's busy on the airport road, and steer clear of the golf course. More at 12."

66 A giant sunshade...

could reflect the Sun's heat.

For many years, some scientists have toyed with the idea of creating a physical barrier between the Earth and the Sun, to shield the planet from extreme heat. This type of project is known as **climate engineering**. It's ambitious, expensive – and controversial...

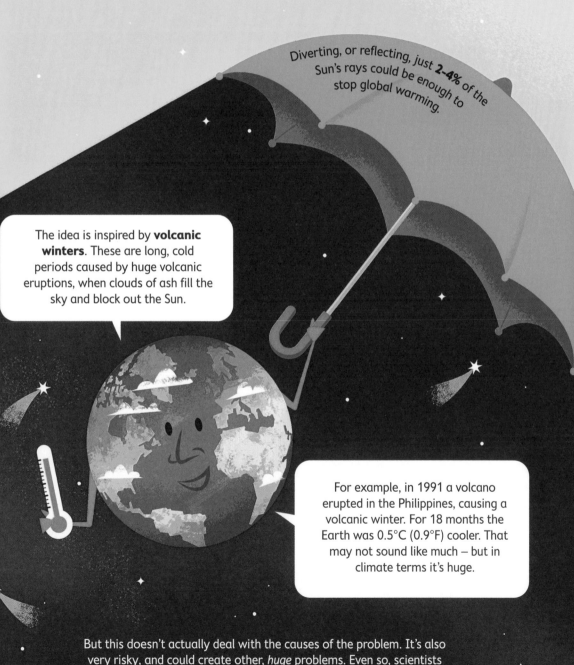

Diverting, or reflecting, just **2-4%** of the Sun's rays could be enough to stop global warming.

The idea is inspired by **volcanic winters**. These are long, cold periods caused by huge volcanic eruptions, when clouds of ash fill the sky and block out the Sun.

For example, in 1991 a volcano erupted in the Philippines, causing a volcanic winter. For 18 months the Earth was 0.5°C (0.9°F) cooler. That may not sound like much – but in climate terms it's huge.

But this doesn't actually deal with the causes of the problem. It's also very risky, and could create other, *huge* problems. Even so, scientists have to explore all sorts of ideas in order to find practical solutions.

67 Forests under water...

can store more carbon than forests on land.

An enormous type of seaweed called **kelp** grows in great swaying forests under the sea. Scientists have discovered it can take in CO_2 and store carbon in colossal amounts.

Kelp sucks up CO_2 dissolved in water and uses it to grow. The CO_2 then remains captured inside the kelp.

Kelp can grow up to **50m (165ft)** tall.

It also grows very quickly — up to **30cm (12 inches)** every day.

Underwater forests like this are known as **blue carbon ecosystems**.

Kelp forests can store **more** carbon than forests on land, so it's vitally important that these ecosystems are protected.

68 Even clouds...

can become extinct.

There is a common type of cloud, called **stratocumulus**, that forms into low, fluffy blankets.

These blankets of cloud cover large parts of the globe, reflecting lots of sunlight and heat, and helping to keep the planet cool.

But we can't take these clouds for granted. In 100 years, they could be gone.

Some scientists are using computers to predict the climate of the future.

One team has found that, if CO_2 levels continue to rise, eventually stratocumulus clouds won't be able to form at all.

They'll become extinct. And without the shade these clouds provide, the planet will heat up faster than ever before.

Computer predictions are helping scientists make the argument...

...that we need to act NOW to save our planet...

...and its clouds.

69 Every bite of cricket cake...

can save a slice of wilderness.

Protein is an important nutrient that helps your body grow. You can get lots of protein by eating beef – but in the future you may want to try another, greener source of protein: insects called crickets.

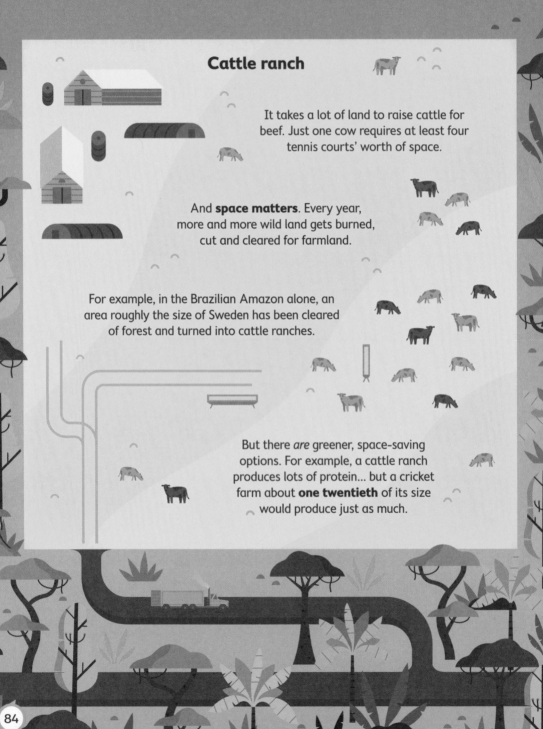

Cattle ranch

It takes a lot of land to raise cattle for beef. Just one cow requires at least four tennis courts' worth of space.

And **space matters**. Every year, more and more wild land gets burned, cut and cleared for farmland.

For example, in the Brazilian Amazon alone, an area roughly the size of Sweden has been cleared of forest and turned into cattle ranches.

But there *are* greener, space-saving options. For example, a cattle ranch produces lots of protein... but a cricket farm about **one twentieth** of its size would produce just as much.

Wilderness

Crickets can be eaten whole, ground into flour for cakes and crackers or made into energy bars.

In other words, by farming crickets we could preserve far more of the planet's wilderness and wildlife, while getting exactly the same amount of protein.

Cricket farm

But crickets aren't the only option. Peas and beans are a good source of protein, and also take up much less land than cattle.

70 Number-crunching farmers...

can help feed the world.

Scientists estimate that to feed the world's growing population for the next 40 years, farmers need to produce *more* food than has *ever* been harvested. But how? Farmers in the Netherlands have an answer: data.

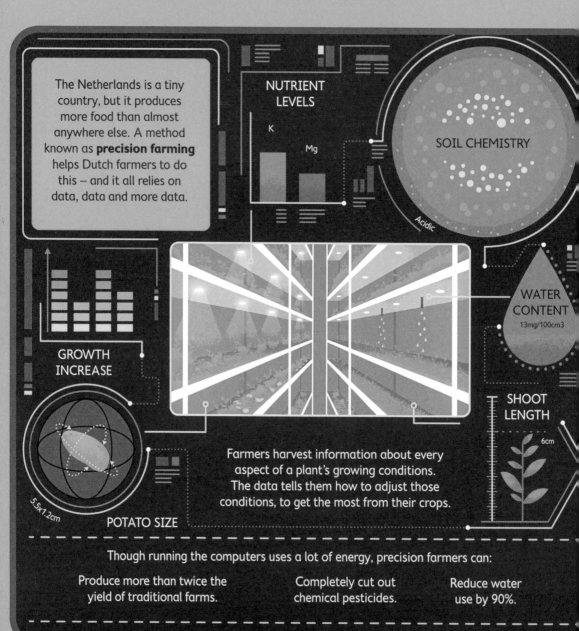

The Netherlands is a tiny country, but it produces more food than almost anywhere else. A method known as **precision farming** helps Dutch farmers to do this – and it all relies on data, data and more data.

NUTRIENT LEVELS

K

Mg

SOIL CHEMISTRY

Acidic

WATER CONTENT
13mg/100cm3

GROWTH INCREASE

SHOOT LENGTH

6cm

5.5x1.2cm

POTATO SIZE

Farmers harvest information about every aspect of a plant's growing conditions. The data tells them how to adjust those conditions, to get the most from their crops.

Though running the computers uses a lot of energy, precision farmers can:

Produce more than twice the yield of traditional farms.

Completely cut out chemical pesticides.

Reduce water use by 90%.

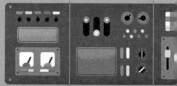

71 Old coal mines...

could store energy like batteries.

In the right conditions, wind turbines and solar panels can generate lots of energy. One group of engineers are developing a way to store some of that energy for later, in disused coal mines that would act like giant batteries...

When the Sun's shining and the wind's blowing, spare **renewable energy** would be used to power motors which lift big weights to the top of a mine shaft.

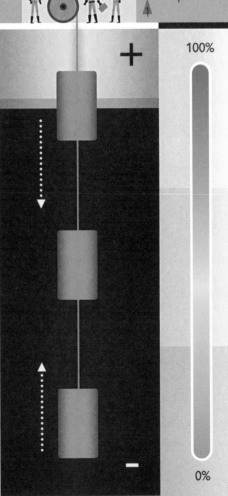

100%

When the wind drops or the Sun goes down, but lots of electricity is needed, the weights would be released to the bottom of the shaft.

The weights are attached to a cable which turns a generator as they are lowered down the shaft. This produces electricity to meet the demand.

The weights could be released quickly to produce a rapid surge of electricity, or more slowly to generate power over a longer period.

The weights can then be hauled back to the top to repeat the process.

0%

Unlike traditional batteries that become less efficient over time, this process could be used to store and discharge large amounts of energy over and over again.

72 Armies of ducks...

protect farmers' fields.

Farmers often use poisonous chemicals known as **pesticides** to kill off insects that feed on their crops. But an ancient, less toxic, technique is making a comeback – sending in ducks.

Farmers take their flocks of ducks into the fields each day until harvest time.

Quack quack all clear in sector 7...

The ducks feast on pests and devour weeds that would otherwise choke the crops.

73 Scientists travel through time...

using old ships' logbooks.

Satellites and weather stations record climate data. But to understand long-term changes in our climate, scientists need to find data from the time before satellite technology existed.

In the 1800s and early 1900s, sailors kept extremely detailed records of the weather they experienced at sea.

Daily records of air pressure, temperature, wind speed and direction, as well as sea ice, were written in a **logbook**, alongside details of the ship's location.

October 10th 1879
Location: 71° 43' 25" N 176° 12' 15" W

Wind direction: south
Wind force: 6
Air temperature: -14°C (6°F)
Surface water temperature: 0°C (32°F)
Ship surrounded by ice.

As the ducks wander through the field, they stir up the soil, releasing nutrients that feed the crops.

Rice farmers in a number of Asian countries have been using ducks in this way for hundreds of years – but farmers elsewhere are starting to use ducks too.

QUAAAAACK locusts spotted in sector 4!

Quack quack, received loud and clear – locust gobblers are on our way.

Today, **citizen scientists** – people who volunteer to help with scientific research – copy out old logbooks...

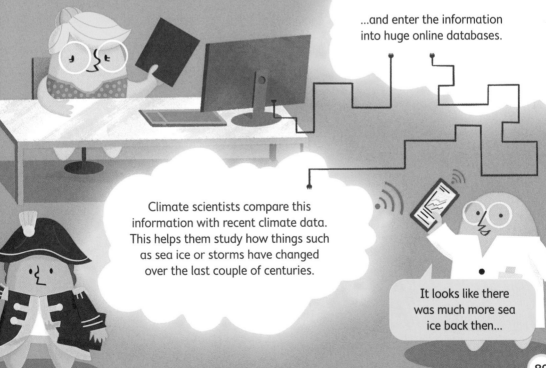

...and enter the information into huge online databases.

Climate scientists compare this information with recent climate data. This helps them study how things such as sea ice or storms have changed over the last couple of centuries.

It looks like there was much more sea ice back then...

74 To rebuild a reef...

you can fake it 'til you make it.

Damaged coral reefs are often abandoned by the creatures that live on them. This makes it hard for these reefs to recover. Scientists may have found a way to trick wildlife into thinking a reef is healthier than it is.

A HEALTHY REEF is full of noise. Thousands of creatures can live on it, making different sounds.

MUNCH

Reef fish munch, grunt and even whoop. Shrimp snap their claws.

SNAP

These sounds travel for miles, attracting new creatures to the reef.

CRUNCH

Healthy reefs are full of food and places to hide.

POP

SNAP

Corals feed on nutrients in fish droppings.

GRUNT

A DAMAGED REEF is quieter – so new fish are much less likely to come.

Scientists hope that damaged reefs will recover more quickly if they can attract young fish – and their nutritious droppings.

WHOOP

CRACKLE

Experiments have shown that playing a recording of a healthy reef attracts **TWICE** as many young fish to live there.

BUZZ

75 The "Doomsday" Vault...
guards dinner from disaster.

On the Arctic island of Svalbard, millions of seeds are stored beneath the ice in the world's largest seed bank – the Svalbard Global Seed Vault. Nicknamed the Doomsday Vault, it stores the seeds of some of the world's most important crops – as a back-up in case of crisis.

The Doomsday Vault was created by the Norwegian government, but seeds have been donated from around the world.

The vault has space to store over **2 billion** seeds.

If crops and regional seed stores are destroyed in a future crisis, samples of seeds can be withdrawn from the vault at Svalbard to help build up food supplies again.

Many modern farms only grow one type of plant – a method of farming known as **monoculture**. This means a single disaster could potentially wipe that crop out.

A monsoon destroyed all our rice fields – we need seeds!

I'm withdrawing wheat seeds. A drought has left my country with no food.

A new disease wiped out our lentil crop, so we're taking out our seeds.

War in our region has destroyed many farms. We need help to start again.

76 One person's passion...

can inspire a movement of millions.

Can you match these individuals with their planet-saving actions, and the impact they had?

I'm concerned about these piles of plastic waste around my village.

Landowners are cutting down the rainforest to make way for cattle farms. In contrast, we local people *look after* the forest, when harvesting rubber or nuts.

CHICO MENDES
1970s, Brazil

I'm worried that big companies have control over the seeds that farmers need to grow crops.

ISATOU CEESAY
1990s, Gambia

VANDANA SHIVA
1980s, India

I'm angry that governments aren't doing more to tackle the climate crisis.

GRETA THUNBERG
2018, Sweden

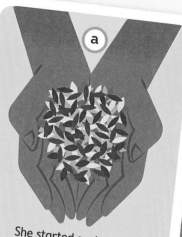

a

She started saving the seeds of traditional local crops, and inspired others to do the same.

b

She taught herself how to make a coin purse out of old plastic bags, then shared her skills with other women.

c

He gathered rubber workers together to block the path of bulldozers.

d

She refused to go to school.

SKOLSTREJK FÖR KLIMATET

1

Within a year, millions of young people around the world had joined her protest.

PROTECT OUR FUTURE ACT NOW SCHOOL STRIKE

2

The rubber workers persuaded Brazil's government to create more than 60 protected areas of rainforest.

3

4,000 different types of rice have been stored safely in seed banks, which farmers can freely access.

4

Hundreds of Gambian women now gather up plastic waste and make a living from making recycled products.

Answers = Chico: c2; Isatou: b4, Vandana: a3; Greta: d1.

77 A ship trapped in ice...

makes an excellent research base.

Every winter, vast regions of the Arctic Ocean freeze, forming a layer of **sea ice** on the surface. Since the 1970s, the amount of sea ice has roughly halved, because the Arctic is getting warmer. To understand why, scientists froze a ship into the ice to use as a laboratory for a year.

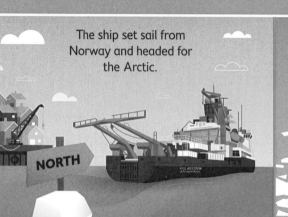

The ship set sail from Norway and headed for the Arctic.

NORTH

Once far enough north to reach icy waters, the ship stopped. As the sea froze over, the ship became locked into the ice.

We're stuck!

Ice eventually surrounded the ship entirely. It froze over an area bigger than a large town.

The researchers lived on the ship and stored equipment and supplies there.

They set up camps on the ice, for conducting research.

We're researching why the Arctic is warming.

We're looking at how the ocean's chemistry is changing.

NEXT STOP:
Wait and see...

Sea ice drifts constantly, and anything caught in it drifts with it. When the crew first lodged the ship into the ice, they didn't know exactly where they would end up.

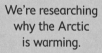

78 Hungry goats...
can tame wildfires.

As the Earth heats up, more wildfires are raging out of control, killing wildlife and destroying homes. To prevent a wildfire from spreading, a gap in vegetation, called a **firebreak**, is often cleared. Enter the goat...

To prepare for potential wildfires, goats can be set loose to munch through grasses, shrubs and bushes.

So if a fire does start, it is less likely to spread. The goats have eaten up a lot of the fuel that would have kept it burning.

But some wildfires are too big for goats to handle on their own. Sometimes firefighters carefully start a fire to clear away vegetation. This is called a **planned burn**.

Another way to create a firebreak is by cutting down trees.

It will become increasingly difficult to control wildfires, unless the *reason* for our hotter, drier weather is tackled: the climate crisis.

Going to space...

makes people care more about Earth.

From a spaceship, the Earth looks like a swirly blue ball that glows against the blackness of space: full of life, beautiful, but fragile too. Many astronauts say that seeing the Earth from afar was a life-changing experience, which made them more committed to saving the planet.

The shift in thinking that astronauts report is known as the **Overview Effect**.

Astronauts can see the whole picture – both the beauty and the damage.

Wildfires

Air pollution

Melting sea ice

Chopped down forests

Actually going to space creates huge CO_2 emissions. But a Dutch former astronaut named **André Kuipers** runs a project which enables children to go on a **Virtual Reality** (VR) spaceflight.

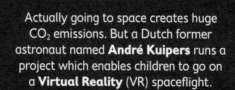

Wearing a VR headset, people see and hear just what an astronaut would. The aim is for them to experience the Overview Effect, without ever leaving the ground.

80 Fly higher or lower...

for clearer, cooler skies.

As planes fly, they often create long ribbons of clouds in the sky behind them. These clouds can cause just as much warming as the fuel the planes burn. Luckily, a solution can usually be found nearby.

When hot air from a plane's engine hits the cold atmosphere, it can create plumes of ice crystals called **contrails**.

These thin clouds can last for up to **18 hours**. By day and by night, they reflect back the heat emitted by the Earth, trapping it near the surface.

Humid air

Drier air

With tens of thousands of flights every day, this heating effect quickly adds up. But just a small adjustment could help.

Most contrails form when planes fly through narrow bands of **humid air**. So, if they can change the height at which they fly, pilots can avoid making more clouds and heating the planet.

Of course, the best thing for the environment is to avoid flying whenever possible.

81 Just enough signatures...

might be all the planet needs.

Some of the most effective planet-saving changes have happened after many governments signed **treaties**. These are agreements that they will follow new rules, and punish those who break the rules.

"We, the undersigned, promise to abide by the terms of the following international treaties..."

CITES TREATY (1973) signed by **183** countries so far

DON'T buy or sell any endangered species of plants or animals.
DO stop trading with any nation that tries to sell these things.

RESULT: up to 35,000 species that were nearly extinct are now protected.

MARPOL (1983) signed by **156** countries so far

DO build ships with separate tanks for oil and waste water.
DON'T empty oil tanks into the sea.
DON'T buy or sell ships that fall below this new standard.

RESULT: Over 95% of all ships meet the latest standards, reducing the amount of pollution in the sea.

The Montreal Protocol (1986) signed by **197** countries so far

DON'T produce or sell products that use polluting gases called CFCs.
DO stop trading with any country that tries to sell CFCs.

RESULT: CFC production has almost entirely stopped. As a direct result, a hole in part of Earth's atmosphere, known as the Ozone Layer, was able to repair itself.

Over 750 treaties concerned with saving the planet have been created, but only a few have been signed by lots of countries. Why don't more sign? It's a hard job crafting a treaty that everyone thinks is fair and achievable — people who manage it are planet-saving heroes.

82 Plastic-eating bacteria...

are powered by bottles.

About a million plastic bottles are sold around the world every minute – and 90% of them are thrown away. Even recycling old bottles to produce new ones releases lots of CO_2 into the atmosphere. But there are bacteria that might help make the process a lot less harmful.

In 2016, a new species of bacteria – *Ideonella sakaiensis* – was discovered in a recycling plant in Japan.

These bacteria actually feed on PET, a plastic that is used to make bottles.

The bacteria break down the plastic. They use carbon from the PET as an energy source and leave behind the raw materials needed to make more PET.

Plastic Sushi restaurant

Scientists are investigating ways to use *Ideonella sakaiensis* on a large scale. This could greatly reduce the amount of CO_2 produced in the process of recycling PET.

83 Skyscrapers of the future...

could be made of wood.

Lots of modern high-rise buildings are made of concrete and steel. But producing these materials can result in massive carbon emissions. That's why some architects are returning to an ancient building material – wood.

Normal wood isn't strong enough, but skyscrapers can be built from specially engineered timber. It's made by sticking layers of wood together so that the grain in each layer runs at right angles to the next, making it extremely strong.

Timber buildings are light and strong, so don't need massive concrete foundations or steel frames. They can withstand powerful earthquakes and temperatures up to **270°C (520°F)** – at which concrete and steel would begin to break down.

There's another benefit too. Trees store carbon in their wood while they are growing (see page 34). If that wood is then used in construction, the carbon will remain locked away for decades.

84 Old Christmas trees...

find new life in January.

What happens to Christmas trees in January, when the strings of lights and ornaments have been put away? Instead of throwing them out, some people use their trees to protect animals and preserve fragile ecosystems.

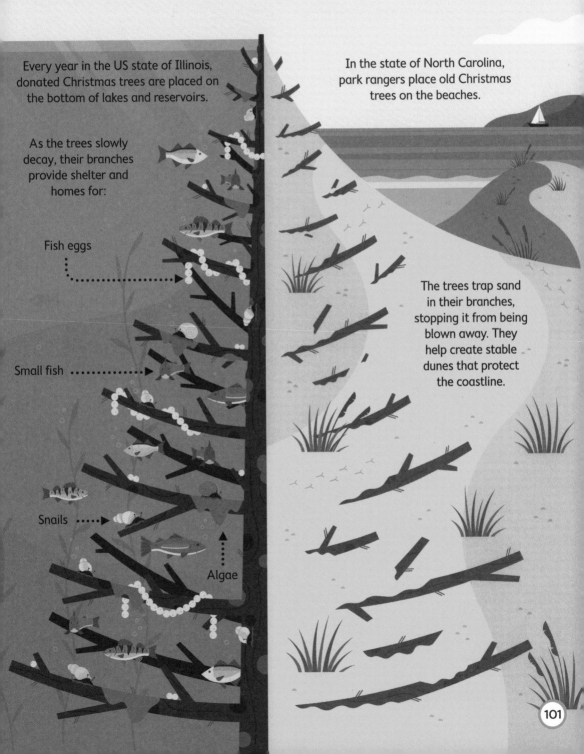

Every year in the US state of Illinois, donated Christmas trees are placed on the bottom of lakes and reservoirs.

As the trees slowly decay, their branches provide shelter and homes for:

Fish eggs

Small fish

Snails

Algae

In the state of North Carolina, park rangers place old Christmas trees on the beaches.

The trees trap sand in their branches, stopping it from being blown away. They help create stable dunes that protect the coastline.

85 A lynx can be found...

in a sample of snow.

Some animals, such as the Canada lynx, can be difficult to track in the wild, making them hard to protect. But something called **environmental DNA**, or **eDNA**, means scientists don't need to see an animal to know it's there...

Canada lynx are solitary, hunt at night, and like to hide from humans, so sightings are extremely rare.

Brrrrrr

It's like looking for a needle in a haystack...

But all animals leave behind traces of their DNA — a genetic code that's unique to each individual — in their urine, droppings, skin cells and hair.

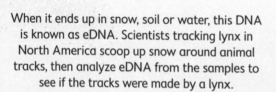

When it ends up in snow, soil or water, this DNA is known as eDNA. Scientists tracking lynx in North America scoop up snow around animal tracks, then analyze eDNA from the samples to see if the tracks were made by a lynx.

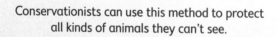

Conservationists can use this method to protect all kinds of animals they can't see.

Stop! A lynx lives here.

It also minimizes the amount of stress caused to the animals — because they don't come into contact with the humans tracking them.

86 Crushing clover...

is the best way to preserve it.

Running buffalo clover is a rare plant found in just a few places in North America. It was in danger of becoming extinct. But then conservationists saw that to save *some* species, you need to get a little creative – and a little destructive.

Running buffalo clover got its name because it grows best in soil churned up and trampled by herds of wild buffalo. But in the 1900s, hunters wiped out the buffalo herds – so the clover nearly died out, too. Luckily, some conservationists figured out how to bring it back.

They used big vehicles to imitate the effect of buffalo hooves. By running over and churning up beds of clover every few years, conservationists are now saving this plant from extinction.

87 Mega turbines...

are 75 floors tall.

The larger the wind turbine, the more of the wind's energy it can convert into electricity. So, over the past few decades, engineers have designed wind turbines to be bigger and BIGGER.

Longer blades on a turbine allows more power to be generated by each rotation.

Back in 1990, the biggest wind turbine was **50m** (164ft) tall. By 2017, the biggest turbine was **246m** (807ft) tall. That's the size of a skyscraper 75 floors tall.

Skyscraper

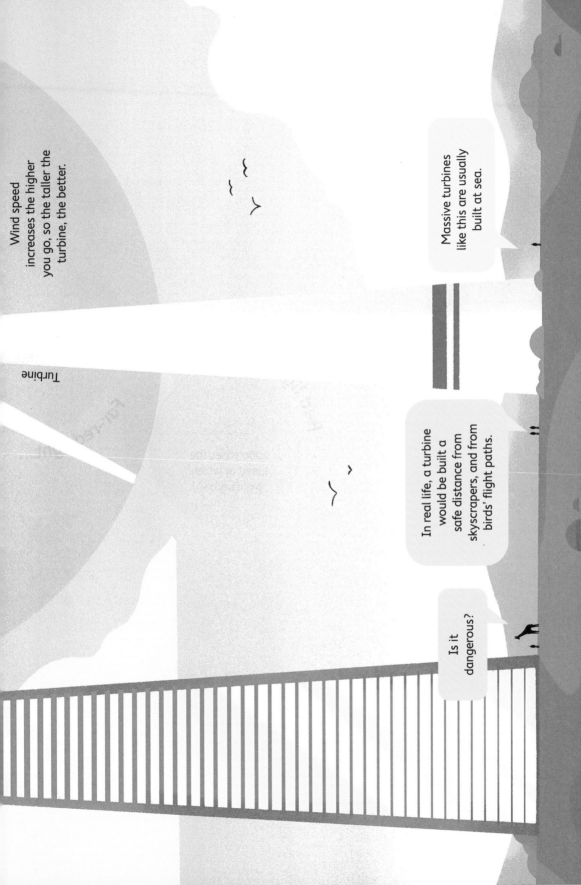

88 Shades of light...

can make plants grow better.

As the world's population grows, more and more land is being cleared for farming. To use up less land, farmers are looking for better ways to grow crops. One way is to use different combinations of light, known as **light recipes**...

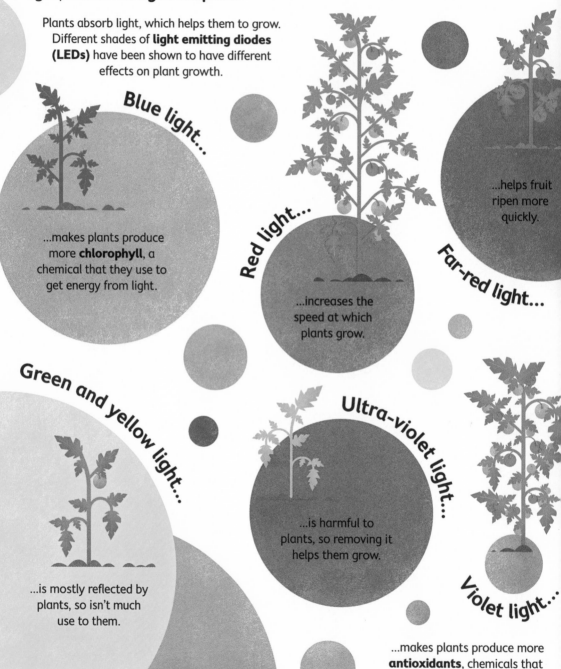

Plants absorb light, which helps them to grow. Different shades of **light emitting diodes (LEDs)** have been shown to have different effects on plant growth.

Blue light...

...makes plants produce more **chlorophyll**, a chemical that they use to get energy from light.

Red light...

...increases the speed at which plants grow.

Far-red light...

...helps fruit ripen more quickly.

Green and yellow light...

...is mostly reflected by plants, so isn't much use to them.

Ultra-violet light...

...is harmful to plants, so removing it helps them grow.

Violet light...

...makes plants produce more **antioxidants**, chemicals that protect them from damage.

89 A massive vacuum cleaner...

sucks smog from the sky.

Cars and factories pollute the air by pumping out harmful gases and specks of dirt, which can make people ill. But a Dutch company has created a tower-shaped vacuum cleaner, which can actually clean the dirty air, or **smog**, around it.

Smog is sucked in.

Tiny specks of dirt stick to the inside of the tower.

Clean air is blown out.

This is the best picnic spot in this city.

When the tower is full of dirt, it is cleaned out.

Within **20m (66ft)** of the tower, the air is up to **70% cleaner**.

Similar technology is being developed for use on a new kind of bike.

A device fitted to the handlebars sucks dirty air in...

...and blows clean air into the rider's face.

90 Sails, kites and bubbles...
could help cut air pollution around the world.

Almost every cargo ship uses an engine powered by heavy diesel fuel, which pumps out polluting gases such as CO_2 and sulfur.
But it doesn't HAVE to be that way.

There are more than **53,000** merchant ships
– ships that carry cargo or passengers –
operating around the world.

Together, they produce up to **5%**
of yearly global CO_2 emissions.
That's even more than planes.

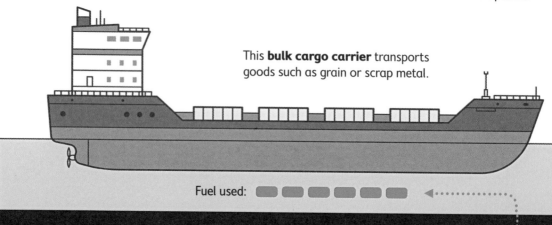

This **bulk cargo carrier** transports
goods such as grain or scrap metal.

Fuel used:

**What could the owners of this cargo ship do to reduce
the total amount of fuel it burns on its voyages?**

HERE ARE A FEW OPTIONS:

REPLACE IT

Build a narrower ship
Longer, narrower ships cut through the
water more easily, and require less fuel.

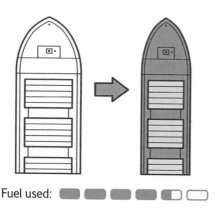

Fuel used:

Build a sailing ship
Cargo ships can be built with stiff sails,
shaped like the wings of a plane.
They are guided by computers and
assist the engine when the wind is fair.

Fuel used:

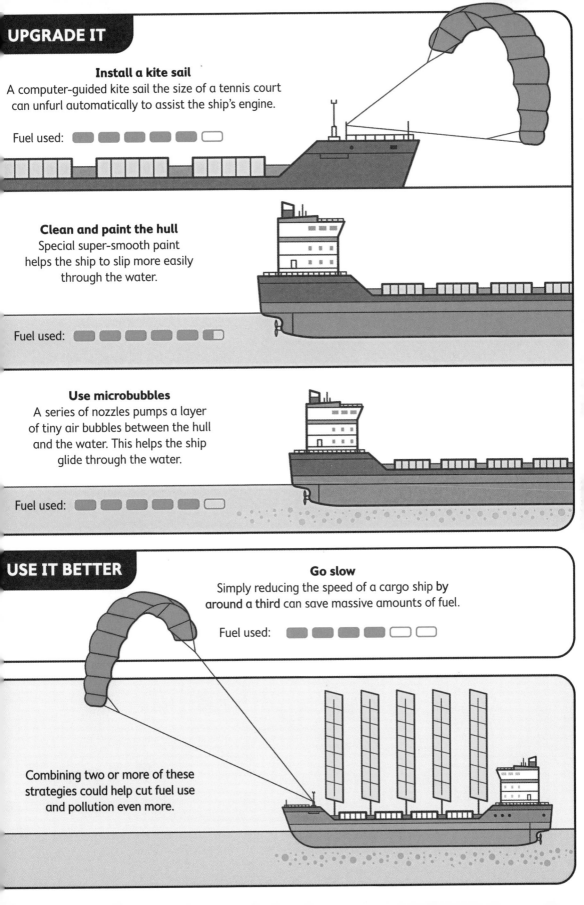

UPGRADE IT

Install a kite sail
A computer-guided kite sail the size of a tennis court can unfurl automatically to assist the ship's engine.

Fuel used:

Clean and paint the hull
Special super-smooth paint helps the ship to slip more easily through the water.

Fuel used:

Use microbubbles
A series of nozzles pumps a layer of tiny air bubbles between the hull and the water. This helps the ship glide through the water.

Fuel used:

USE IT BETTER

Go slow
Simply reducing the speed of a cargo ship by around a third can save massive amounts of fuel.

Fuel used:

Combining two or more of these strategies could help cut fuel use and pollution even more.

91 Killing a rat...

could save a reef.

Coral reefs are under threat in oceans around the world. But there *may* be a way to save some reefs from destruction – without even leaving dry land.

Scientists have found that one key to healthy reefs is having islands nearby with lots of seabirds living on them.

But when humans started settling on islands, they brought along the seabirds' worst enemy: rats.

Seabirds fertilize the islands with their rich, smelly droppings.

Rats devour birds and their eggs, ravaging whole populations.

Nutrients from bird droppings wash down into the water.

With far fewer birds, far fewer nutrients wash into the ocean.

The nutrients feed seaweed, sponges, algae and all the fish that keep a coral reef healthy.

This leads to far fewer fish, and weaker, less healthy coral reefs.

One study showed that islands without rats have **760 times more birds** than rat-infested islands, **50% more fish**, and much more, much **healthier coral**. Using traps and poisons, people have succeeded in wiping out rats on more than 550 islands around the world.

92 Killing rats is always wrong...

or *is* it?

Humans have caused all kinds of environmental problems. Trying to fix these problems is rarely easy – and seemingly simple solutions can bring up complex questions and moral dilemmas. Here is just one example:

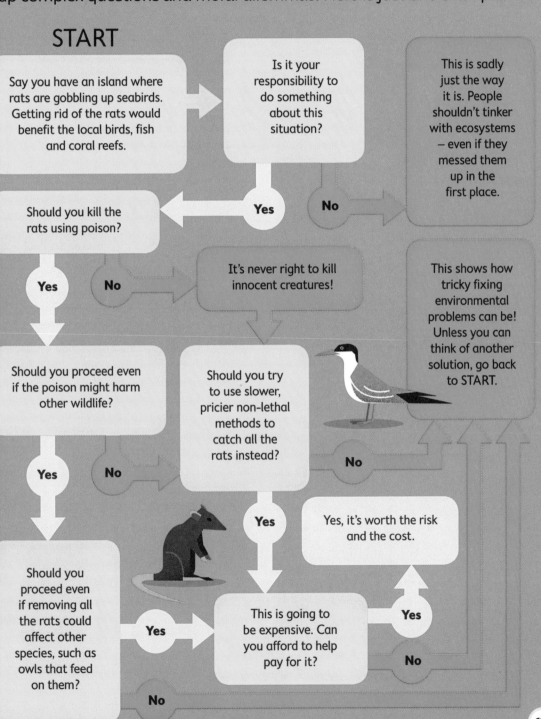

START

Say you have an island where rats are gobbling up seabirds. Getting rid of the rats would benefit the local birds, fish and coral reefs.

Is it your responsibility to do something about this situation?

No → This is sadly just the way it is. People shouldn't tinker with ecosystems – even if they messed them up in the first place.

Yes → Should you kill the rats using poison?

No → It's never right to kill innocent creatures!

Yes → Should you proceed even if the poison might harm other wildlife?

Should you try to use slower, pricier non-lethal methods to catch all the rats instead?

This shows how tricky fixing environmental problems can be! Unless you can think of another solution, go back to START.

No

Yes

Yes → Yes, it's worth the risk and the cost.

Yes → Should you proceed even if removing all the rats could affect other species, such as owls that feed on them?

Yes → This is going to be expensive. Can you afford to help pay for it?

No

No

111

93 Lawns of green, green grass...

aren't very "green" at all.

Beautifully mown lawns are a common sight across the world. But maintaining these spaces can do more harm than good. Adding a little wilderness back to green spaces is much better for the environment.

Mowing lawns can destroy insect nests and animal homes.

Lawnmowers use fuel and pump harmful CO_2 into the atmosphere.

Keeping a lawn green often requires huge amounts of water and chemical fertilizers.

Fertilizers harm wildlife, and break down into **nitrous oxide** – a dangerous greenhouse gas.

Instead, lawn space could be given over to local wildflowers, shrubs and trees, which grow more easily without fertilizers, and need less watering.

The plants provide homes and food for animals and insects.

94 Doing nothing...

can bring wildlife back to outdoor spaces.

Stop! Put away the shovel, rake and shears, and have a nap instead. *Neglecting* outdoor spaces such as gardens and parks can actually increase the amount of food and shelter available to wild animals.

Don't clear away that pile of rotting logs – we'll make our homes in the cool, damp wood.

Don't prune those dried up flowers – we'll build our nests in their hollow stems...

...And I'll snack on their seeds all winter long.

Don't chop down those stinging nettles – I'll nibble their leaves before turning into a butterfly.

Don't fix that hole at the bottom of the fence – that's how I get around from place to place.

Don't throw out the pile of leaves in the corner – I'll spend the winter sleeping snugly under there.

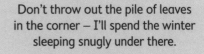

95 Bees and trees...

create work for former coal miners.

Saving the planet means stopping using coal as fuel. So what happens to people who lose their jobs when coal mines close? One option is to find work as beekeepers...

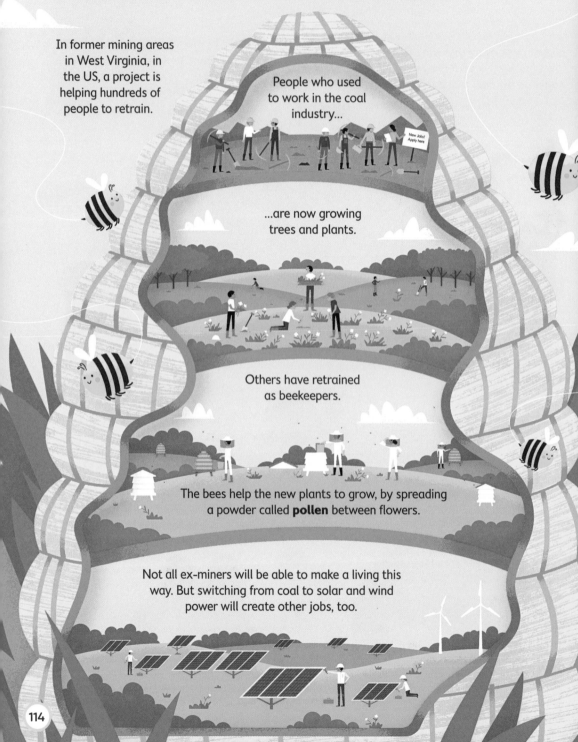

In former mining areas in West Virginia, in the US, a project is helping hundreds of people to retrain.

People who used to work in the coal industry...

New Jobs! Apply here

...are now growing trees and plants.

Others have retrained as beekeepers.

The bees help the new plants to grow, by spreading a powder called **pollen** between flowers.

Not all ex-miners will be able to make a living this way. But switching from coal to solar and wind power will create other jobs, too.

96 Hugging trees...

really did help save a forest.

In March 1974, a team of loggers arrived on the edge of a forest in Chamoli, northern India. Four days later, they left, promising not to return, after local villagers blocked the work by literally hugging the trees.

We've suffered landslides and floods over the last four years, all because the trees that used to hold the ground steady have been chopped down.

Hands off our trees!

In 1980, after years of protests. the Prime Minister of India banned tree-felling in the area for 15 years, to allow the forests to grow back.

We need these trees to keep our village safe.

The protest was publicized throughout India, and around the world. Ever since, tree-huggers in all sorts of places have helped to protect forests.

97 Cities of the future...

will have "smart" windows and "kinetic" paving.

Cities are responsible for a lot of wasted energy, as well as up to **70%** of CO_2 emissions. But scientists and city planners are already developing many ways to make cities cleaner and more energy efficient. Here are just a few...

Green roofs and **green walls** are gardens planted on rooftops and up the sides of buildings.

The plants absorb CO_2 from the air. They even provide soundproofing, and insulate the buildings, too.

Windows can be made from a material called **smart glass**. It responds to the weather and traps heat in a building, or allows it to escape, depending on what's needed.

Cars can be adapted to run on used **cooking oil**. This produces less pollution than fossil fuels and reduces food waste.

Electric buses produce fewer greenhouse gases than buses with standard engines.

Kinetic paving is designed to generate electricity from people's footsteps as they walk along.

Cool roofs use materials designed to reflect sunlight, which reduces the need for air conditioning.

Collecting rainwater reduces the amount of water that needs to be pumped into the city, a process which uses a lot of energy.

Parks filled with trees and plants help to absorb lots of CO_2.

Keeping buildings close together means goods are transported across shorter distances, and people can walk or cycle instead of driving.

The brakes of trains also produce heat, which can be collected and used elsewhere.

Heat energy can be collected in underground train stations from people walking, talking and breathing.

98 A burning river...

fed fuel to an environmental movement.

By the 1960s, people across the United States had started to notice the harm that industrial pollution was causing. After several environmental disasters – including a river on fire – they began to demand change.

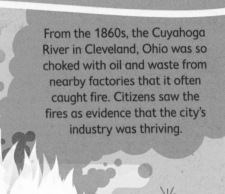

From the 1860s, the Cuyahoga River in Cleveland, Ohio was so choked with oil and waste from nearby factories that it often caught fire. Citizens saw the fires as evidence that the city's industry was thriving.

In 1969, the river caught fire again. Around the same time, a number of other environmental disasters struck across the US.

NO TO POLLUTION

CLEAN OUR RIVER

People staged protests demanding action from the government.

In 1970, the US government founded the Environmental Protection Agency (EPA).

The EPA creates laws to protect people's health and the natural environment, by restricting what chemicals and pollutants industries can use.

Today, much of the Cuyahoga River is clean and filled with wildlife.

99 Tree detectives...

fight illegal logging.

To keep forests healthy, laws exist to prevent loggers from chopping down certain kinds of trees. But these laws are often broken. So there are tree detectives who use science to crack down on the criminals.

Wood from legally cut trees comes with a certificate, but sometimes the paperwork is faked. So what's needed is a **DNA test**.

Every living thing has different DNA: a chemical containing a code that tells it how to grow.

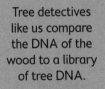

Tree detectives like us compare the DNA of the wood to a library of tree DNA.

From that, we can identify the type of tree and the location of the forest where it was chopped down.

Call the police! This is rosewood from Madagascar. It is illegal to cut it down and sell it abroad.

LEGAL & CERTIFIED

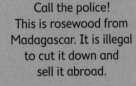

NEW IN

100 If humans can land on the Moon...

why can't they solve the climate crisis?

It once seemed completely far-fetched that a person could ever walk on the Moon. But, in 1969, it happened. Now we are facing another huge challenge: how to solve the climate crisis. Some environmentalists and economists think that governments need to apply the same problem-solving that took astronauts to the Moon. This is known as **Moonshot Thinking.**

Here's what could happen if Moonshot Thinking were applied to solving the climate crisis...

All world leaders would inspire people to feel proud of saving the climate.

AN INSPIRING LEADER

Here are the elements that combined to make the Moon landing happen in July, 1969...

In 1961, US President John F. Kennedy, got Americans excited about the Moon landing project.

Governments would race to cut their country's CO_2 emissions down to zero.

Governments would invest huge sums of money in less harmful energy, buildings and transportation. They would stop spending any money on fossil fuels.

There would be a massive surge in inventions of new and improved technology, such as better solar panels, wind turbines and electric vehicles.

Seeing action at a government level, billions of people would also make changes in their own lives to cut CO_2 emissions.

Right now, *some* of this is happening in *some* places. But Moonshot Thinking could make it happen on a WIDER scale and much more QUICKLY.

URGENCY

MONEY

SCIENTISTS AND ENGINEERS

PUBLIC PASSION

The United States was racing to land on the Moon before anyone else did. Kennedy promised that the American Space Agency would get there within the decade.

Billions of dollars were invested in the project.

The best scientists and engineers prioritized the development of new rockets and spaceships.

600 million people around the world watched the Moon landing on TV.

10 everyday actions...
that will help save the planet.

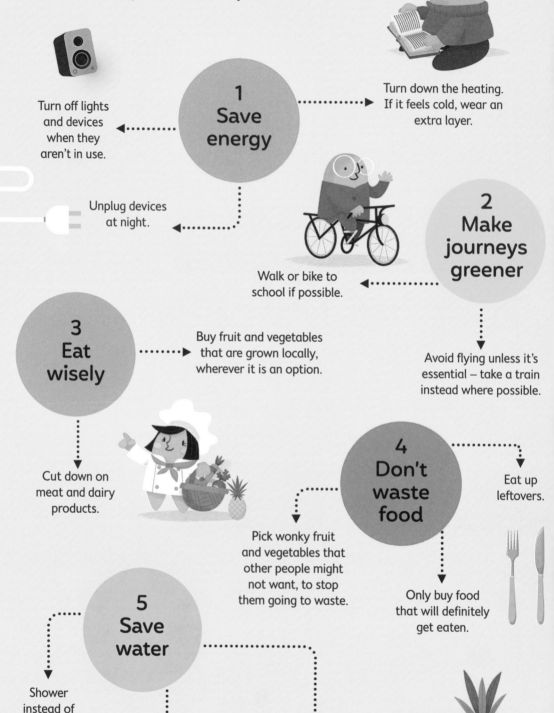

1 Save energy

Turn off lights and devices when they aren't in use.

Turn down the heating. If it feels cold, wear an extra layer.

Unplug devices at night.

2 Make journeys greener

Walk or bike to school if possible.

Avoid flying unless it's essential – take a train instead where possible.

3 Eat wisely

Buy fruit and vegetables that are grown locally, wherever it is an option.

Cut down on meat and dairy products.

4 Don't waste food

Eat up leftovers.

Pick wonky fruit and vegetables that other people might not want, to stop them going to waste.

Only buy food that will definitely get eaten.

5 Save water

Shower instead of taking a bath.

Don't leave the water running while brushing teeth.

Collect rainwater to water plants.

6 Care for nature

Plant wildflowers to help bees thrive.

Support local wildlife charities. Some need help restoring habitats or counting birds and bugs.

Leave gardens and outdoor spaces slightly wild, to boost biodiversity.

7 Change shopping habits

Avoid buying things which come wrapped in lots of packaging.

Think twice before buying something new.

When buying something new, pick a thing that can be used again and again.

Hire things if possible, buy second hand, or exchange with friends.

8 Throw less stuff away

Try to repair things that are broken.

Say no to single-use plastic bags, straws, cups, knives and forks. Use reusable ones instead.

Before chucking something out, think about who might want it. Donate to charity or sell your things second hand.

9 Recycle

If throwing something away, check if it is recyclable and dispose of it correctly.

Depending on where you live, it may be possible to recycle paper, cardboard, glass, metal and different types of plastic.

10 Spread the word

Encourage friends, family and teachers to make planet-saving choices, too.

123

Glossary

This glossary explains some of the words used in this book.
Words written in *italic* type have their own entries.

algae A plant-like type of *organism*, including seaweed, that generally grows in or near water

atmosphere The layer of gases surrounding the Earth

bacteria Microscopic living things, found all over the world

battery A device that stores energy, and powers objects using *electricity*

biodegradable A substance or material that breaks down over time, and is absorbed into the *environment*

biodiversity The variety of plant and animal *species* in any one place

carbon A chemical element that forms many different molecules, and is the building block of life on Earth

carbon footprint The amount of carbon dioxide that is released in the process of making or doing something

carbon storage Taking carbon dioxide out of the *atmosphere*, often by trees or plants which use the *carbon* to grow

climate The typical or average weather conditions in a particular region

climate change The change in Earth's *climate* over time

climate crisis The huge and far-reaching negative effects of *climate change* on the planet

compostable A way to describe a substance or material that *biodegrades* completely, leaving behind nothing toxic at all

conservation The effort to protect and preserve *species* and their *environments*

deforestation The act of chopping or burning down large areas of forest

DNA The chemical code that tells living things how to build themselves

ecosystem A community of plants, animals and their *environment*

electricity A form of energy. It is used to power all sorts of things, from televisions and computers, to lights and trains.

emissions Gases, such as *greenhouse gases*, that are pumped into the *atmosphere*

endangered species A *species* that is at serious risk of *extinction*

environment The surrounding conditions in which living things exist

extinction When the last member of a *species* dies, that species is extinct.

fertilizer A substance that can be added to soils to help plants grow

fossil fuel A *fuel*, such as coal, oil or gas, made from fossilized remains of living things. When burned, fossil fuels produce energy, but also emit *greenhouse gases*.

fuel Something burned to generate energy, particularly in a vehicle

fuel efficiency How well a device or vehicle turns *fuel* into energy. Greater fuel efficiency means less fuel is wasted.

green A common word for something environmentally friendly that doesn't cause lots of damage or *pollution*

greenhouse gas A gas that traps heat in the *atmosphere*. These gases include carbon dioxide and methane, and are particularly responsible for *climate change*.

habitat The place where an animal or plant *species* lives

insulation A substance used to stop heat or sound being lost or wasted

invasive species A *species* introduced to an *environment* that damages local plants and animals

landfill site An area used to dump waste that cannot be *recycled* or *composted*. Waste here can hang around for hundreds of years.

microplastics Tiny pieces of *plastic*, smaller than 5mm (0.2 inches). They seep into the *environment* and harm the animals that eat them.

migration Movement from one place to another. Many animals migrate each season to mate or find food.

mining Digging *fossil fuels* and other minerals out of the ground

monoculture Growing just one crop on a field or farm

national park Also known as a nature reserve or a protected area, this is a place protected by laws, to help conserve the plants and animals that live there.

nutrients Substances or chemicals that provide nourishment, and help an *organism* live and grow

oil spill When oil, a *fossil fuel*, is released into the *environment*, particularly the sea, and damages life there

organism A single living thing, for example a plant, animal, fungus or *bacteria*

plastic A material that can be made into any shape, and usually lasts a very long time

pollution When harmful substances known as pollutants, such as waste, dirt and exhaust fumes, contaminate the *environment*

recycling Turning used objects and materials into new ones

renewable energy Energy generated by a source that never runs out, for example by *solar panels* or *wind turbines*

sea level The average level of the sea's surface where it meets land

single-use plastic A *plastic* object or piece of packaging that is used just once, then thrown away into a *landfill site*

solar panel An energy generator that turns light from the Sun into *electricity* or heat

species A type of plant, animal or other living thing

treaty An agreement signed by two or more countries

water footprint The amount of fresh water it takes to create or produce something

wildfire A fire that burns out of control through forests or grassland

wind turbine An energy generator that turns the movement of the wind into *electricity*

Index

Making this book...

took a team of passionate planet-savers.

Research and writing:
Rose Hall, Jerome Martin, Alice James,
Darran Stobbart, Alex Frith, Tom Mumbray,
Eddie Reynolds, Lan Cook and Matthew Oldham

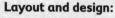

Layout and design:
Jenny Offley, Lenka Hrehova, Samuel Gorham,
Tilly Kitching, Lucy Wain, Winsome D'Abreu,
Freya Harrison and Matt Preston

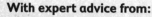

Illustration:
Parko Polo, Dominique Byron,
Dale Edwin Murray, Federico Mariani,
Jake Williams and Ollie Hoff

With expert advice from:
Professor Mike Berners-Lee, Lancaster University
and Small World Consulting
Jessica Moss, Small World Consulting
Professor Owen Lewis, University of Oxford

Series editor: Ruth Brocklehurst
Series designer: Stephen Moncrieff

With thanks to Ananas Anam (Piñatex — pineapple leaf leather alternative, p12), Modern Meadow (collagen protein leather alternative, p12), Zelp (p36), Ecorobotix (p42), Gravitricity (p87), Studio Roosegaarde (p107) and Pavegen (kinetic pavements, p116) for permission to include their products. 'Support' © Lorenzo Quinn, is reproduced on p55 by kind permission of Lorenzo Quinn, illustration by Parko Polo.

First published in 2020 by Usborne Publishing Ltd., Usborne House, 83-85 Saffron Hill, London, EC1N 8RT, United Kingdom, usborne.com Copyright © 2020 Usborne Publishing Ltd. The name Usborne and the Balloon logo are trade marks of Usborne Publishing Ltd. All rights reserved. No part of this publication may be reproduced, stored in any retrieval system, or transmitted in any form or by any means, electronic, mechanical, photocopying, recording or otherwise, without the prior permission of the publisher. Printed in UAE. UE.

Usborne Publishing is not responsible and does not accept liability for the availability or content of any website other than its own, or for any exposure to harmful, offensive or inaccurate material which may appear on the Web. Usborne Publishing will have no liability for any damage or loss caused by viruses that may be downloaded as a result of browsing the sites it recommends.